Lecture Notes on Mathematics
in the Life Sciences

For further volumes:
http://www.springer.com/series/10049

Avner Friedman • Chiu-Yen Kao

Mathematical Modeling
of Biological Processes

 Springer

Avner Friedman
Department of Mathematics
The Ohio State University
Columbus, OH, USA

Chiu-Yen Kao
Department of Mathematical Sciences
Claremont McKenna College
Claremont, CA, USA

ISSN 2193-4789   ISSN 2193-4797 (electronic)
ISBN 978-3-319-08313-1   ISBN 978-3-319-08314-8 (eBook)
DOI 10.1007/978-3-319-08314-8
Springer Cham Heidelberg New York Dordrecht London

Library of Congress Control Number: 2014945230

Mathematics Subject Classification: 35Q92, 37N25, 92C17, 92-01, 97M60, 92C50

Printed on acid-free paper

Springer is part of Springer Science+Business Media (www.springer.com)

Printed by Markono Print Media Pte Ltd

Contents

CHAPTER 1

Introduction

Mathematical biology is characterized by the origin of the problems that it engages: they all come from biology and may borrow from a variety of mathematical disciplines such as ordinary and partial differential equations, probability, statistics, and discrete mathematics. The common aim is to gain better understanding of the biological processes through mathematical ideas and numerical simulations. Biological processes typically involve a large number of interacting components, and it is often a challenge to determine how much each component contributes to the final outcome. This is precisely where a mathematical model of the process may become a useful tool for biologists. It is thus not surprising that, together with the great advances in the biological sciences, mathematical biology has become in recent years a fast-growing field of research within the mathematical sciences.

Historically, applications have been the driving force in the development of science and in mathematics. But applications can also serve to motivate the teaching of mathematics in school. The present book is based on a course developed for graduate students with basic knowledge in ordinary differential equations and some familiarity with partial differential equations and biology, although the biology background is not absolutely required. The aim of the course is to demonstrate to the students the power of mathematics and computational codes in setting up biological processes within a rigorous and predictive framework. At the same time the students will have gained highly motivated knowledge in ordinary and partial differential equations and will have learned how to program with MATLAB without previous programming experience and how to use codes in order to test biological hypotheses.

The book includes a selection of biological topics: enzyme dynamics, spread of disease, harvesting bacteria, competition among live species, neuronal oscillations, transport of neurofilaments in axon, cancer and cancer therapy, and granulomas. Each of these topics begins with a description of the biological background and a biological question that requires the use of mathematics, then proceeds to develop a mathematical model and analysis of the model, and finally introduces a numerical implement that the students will use in order to check, by simulations, various predictions of the model. The book also includes mathematical introduction to ordinary differential

© Springer International Publishing Switzerland 2014
A. Friedman, C.-Y. Kao, *Mathematical Modeling of Biological Processes*, Lecture Notes on Mathematical Modelling in the Life Sciences, DOI 10.1007/978-3-319-08314-8_1

equation, to conservation laws, and to diffusion equations. Each chapter includes two sets of exercises: one set relates to the analysis of the model and another set relates to numerical simulations of the model. Chapters 2–7 of the book use only ordinary differential equations, while the remaining chapters use partial differential equations, such as systems of conservation laws and diffusion equations. The book can be used either for a semester course or as a basis for a one-year course.

We found it useful in this course to include "projects" for the students. We divide the class into small groups, and each group is assigned a research paper which they are to present to the entire class at the end of the course. It is also very helpful if the instructors themselves present such a "project" prior to the students' presentations and engage the class in free discussion and critics of the paper presented.

It is our hope that the book will help demonstrate to the students, and to other readers, the challenge and excitement that is currently taking place at the interface of mathematics and biology.

During the preparation of the book, Chuan Xue taught the numerical sections of the book and Ching-Shan Chou taught the numerical sections of an undergraduate version of the book at Department of Mathematics, The Ohio State University; we acknowledge with thanks the many suggestions they have made.

CHAPTER 2

Chemical Kinetics and Enzyme Dynamics

Cells are the basic units of life. A cell consists of a concentrated aqueous solution of molecules contained in a membrane, called **plasma membrane**. A cell is capable of replicating itself by growing and dividing. Cells that have a nucleus are called **eukaryotes**, and cells that do not have a nucleus are call **prokaryotes**. Bacteria are prokaryotes, while yeast and amoebas, as well as most cells in our body, are eukaryotes. The **deoxyribonucleic acid (DNA)** is a very long polymeric molecule, consisting of two strands of chains, having double helix configuration, with repeated nucleotide units A, C, G, and T. The DNA is packed in chromosomes, within the nucleus in eukaryotes. In humans, the number of chromosomes is 46, except in sperm and egg cells where the number is 23.

The DNA is the genetic code of the cell; it codes for proteins. Proteins lie mostly in the cytoplasm of the cells, that is, outside the nucleus; some proteins are attached to the plasma membrane, while some can be found in the nucleus. Proteins are polymers of amino acids whose number typically ranges from hundreds to thousands; there are 20 different amino acids from which all proteins are made. Each protein assumes three-dimensional configuration, called **conformation**. Proteins perform specific tasks by changing their conformation.

Two proteins, A and B, may combine to form a new protein C. We express this process by writing

$$A + B \to C.$$

Biological processes within a cell involve many such reactions. Some of these reactions are very slow, others are very fast, and in some cases the reaction rate may start slow, then speed up until it reaches a maximal level. In this chapter we consider the question: how to determine the speed of biochemical reactions among proteins. In order to address this question we shall develop some mathematical models.

We begin with a simple case. Suppose we have two proteins, A and B, or more generally two molecules A and B. We assume that A and B, when coming in contact, undergo a reaction, at some rate k_1, that makes them form a new molecule C. We express this reaction by writing

© Springer International Publishing Switzerland 2014
A. Friedman, C.-Y. Kao, *Mathematical Modeling of Biological Processes*, Lecture Notes on Mathematical Modelling in the Life Sciences, DOI 10.1007/978-3-319-08314-8_2

$$A + B \xrightarrow{k_1} C;$$

k_1 is called the **rate coefficient**. The respective concentrations of three molecules are denoted by $[A]$, $[B]$, and $[C]$. The **law of mass action** states that the **reaction rate** $\frac{d[C]}{dt}$, or v_1, of the above reaction is given by

$$v_1 = k_1[A][B],$$

that is,

$$\frac{d[C]}{dt} = k_1[A][B].$$

Then also

$$\frac{d[A]}{dt} = -v_1 = -k_1[A][B], \qquad \frac{d[B]}{dt} = -v_1 = -k_1[A][B].$$

Suppose one molecule of A and two molecules of B react to form a new molecule C:

$$A + 2B \xrightarrow{k_1} C. \tag{2.0.1}$$

Since this reaction can be viewed as $A + B + B \to C$, the law of mass action states that

$$\frac{d[C]}{dt} = v_1, \quad \text{where} \quad v_1 = k_1[A][B]^2,$$

and then

$$\frac{d[A]}{dt} = -v_1, \qquad \frac{d[B]}{dt} = -2v_1.$$

The **stoichiometric coefficients** of A, B, C in this reaction are 1, 2, 1. For the reversible reaction

$$C \xrightarrow{k_2} A + 2B, \tag{2.0.2}$$

the reaction rate $\frac{d[C]}{dt}$ is $-v_2$ where $v_2 = k_2[C]$, and under both reactions, (2.0.1) and (2.0.2),

$$\frac{d[A]}{dt} = -v_1 + v_2,$$

$$\frac{d[B]}{dt} = -2v_1 + 2v_2,$$

$$\frac{d[C]}{dt} = v_1 - v_2,$$

or, in a vector form,

$$\frac{d}{dt} \begin{pmatrix} [A] \\ [B] \\ [C] \end{pmatrix} = \begin{pmatrix} -1 & 1 \\ -2 & 2 \\ 1 & -1 \end{pmatrix} \begin{pmatrix} v_1 \\ v_2 \end{pmatrix}.$$

Consider n chemical species whose concentrations are $[X_1]$, $[X_2]$,...,$[X_n]$. Let there be r chemical reactions, with each reaction being symbolized by

$$s_{1j}^R X_1 + s_{2j}^R X_2 + \ldots + s_{nj}^R X_n \to s_{1j}^P X_1 + s_{2j}^P X_2 + \ldots + s_{nj}^P X_n \quad (j = 1, \ldots, r) \tag{2.0.3}$$

where s_{ij}^R, s_{ij}^P are stoichiometric coefficients of species i on the reactant side and product side, respectively, of reaction j. Let $\mathbf{X}$ be the concentration vector $[[X_1], \ldots, [X_n]]$, $\mathbf{v}$ the reaction velocity vector $[v_1, \ldots, v_r]$ where v_j is the rate of reaction j and S the so-called **stoichiometric matrix** whose element s_{ij} is equal to $(s_{ij}^P - s_{ij}^R)$. The general set of dynamics equations for chemical reactions systems can be written succinctly as $\frac{d[X_l]}{dt} = \sum_{j=1}^r s_{lj}v_j$, or

$$\dot{\mathbf{X}} = \mathbf{S}\mathbf{v}, \tag{2.0.4}$$

where $\dot{\mathbf{X}}$ means $d\mathbf{X}/dt \equiv [d[X_1]/dt, \ldots, d[X_n]/dt]$. Such a system of ODEs is called a **stoichiometric dynamics system**.

A reaction j is said to have **mass-action kinetics** if its rate v_j has the form

$$v_j = k_j \Pi_{i=1}^n [X_i]^{s_{ij}^R},$$

and, if this holds for all j, then (2.0.4) can be written in the form

$$\frac{d[X_l]}{dt} = \sum_{j=1}^r s_{lj} k_j \Pi_{i=1}^n [X_i]^{s_{ij}^R} \quad (1 \le l \le n). \tag{2.0.5}$$

In the next chapter we shall review the basic theory of ordinary differential equations. By solving the system of differential equations (2.0.5) we can find how each concentration $[X_l]$ will evolve as a function of time. This is illustrated in Problems 2.1 and 2.2.

Metabolism in a cell is the sum of physical and chemical processes by which material substances are produced, maintained, or destroyed and by which energy is made available. **Enzymes** are proteins that act as catalysts in speeding up chemical reactions within a cell. They play critical roles in many metabolic processes within the cell. An enzyme, say E, can take a molecule S and convert it to a molecule P in one millionth of a second. The original molecule S is referred to as the **substrate**, and P is called the **product**. The enzyme-catalyzed conversion of a substrate S into a product P is written in the form

$$S \xrightarrow{E} P. \tag{2.0.6}$$

The profile $[S] \to [P]$ can take different forms, depending on the underlying biology. Two typical profiles are shown in Fig. 2.1.

Figure 2.1a, b have been shown to hold in different experiments, but it would be useful to derive them by mathematical analysis based on known properties of enzymes. We begin with the derivation of a formula that yields the profile of Fig. 2.1a.

In what follows we show how such a profile can be derived from the law of mass action. We write, schematically,

$$S + E \underset{k_{-1}}{\overset{k_1}{\rightleftharpoons}} C$$

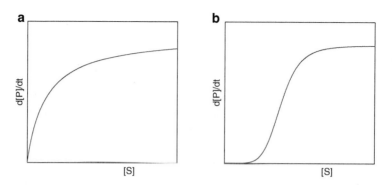

FIGURE 2.1. Two different profiles of the enzymatic conversion of $S \to P$

where C is the complex SE,

$$C \xrightarrow{k_2} E + P.$$

By the law of mass action

$$\frac{d[C]}{dt} = k_1[S][E] - (k_{-1} + k_2)[C], \qquad (2.0.7)$$

$$\frac{d[E]}{dt} = -k_1[S][E] + (k_{-1} + k_2)[C], \qquad (2.0.8)$$

$$\frac{d[S]}{dt} = -k_1[S][E] + k_{-1}[C], \qquad (2.0.9)$$

$$\frac{d[P]}{dt} = k_2[C]. \qquad (2.0.10)$$

Notice that

$$\frac{d}{dt}\left([E] + [C]\right) = 0$$

so that $[E] + [C] = const = e_0$; e_0 is the total concentration of the enzyme in both E and the complex C. Note also that $\frac{d[C]}{dt} + \frac{d[S]}{dt} + \frac{d[P]}{dt} = 0$; hence Eq. (2.0.9) follows from Eqs. (2.0.7) and (2.0.10) and may therefore be dropped.

We focus on Eq. (2.0.7) and note that the enzymatic process is very fast, that is, the rates of formation and breakdown of the complex C are very fast. We assume that these rates are essentially the same, so that $\frac{d[C]}{dt}$ is approximately zero. Hence, approximately,

$$k_1[S][E] - (k_{-1} + k_2)[C] = 0. \qquad (2.0.11)$$

Substituting $[E] = e_0 - [C]$ we get

$$k_1[S](e_0 - [C]) - (k_{-1} + k_2)[C] = 0$$

or

$$[C] = \frac{k_1 e_0 [S]}{(k_{-1} + k_2) + k_1 [S]} = \frac{e_0 [S]}{K_M + [S]}$$

where $K_M = \frac{k_{-1} + k_2}{k_1}$.

Recalling (2.0.10) we have thus derived the **Michaelis–Menten formula**

$$\frac{d[P]}{dt} = \frac{V_{max}[S]}{K_M + [S]} \qquad (2.0.12)$$

where $V_{max} = k_2 e_0$ and K_M are constants; note that

$$\frac{d[P]}{dt} \to V_{max} \quad \text{as} \quad [S] \to \infty.$$

The assumption we made in the derivation of (2.0.12) that $d[C]/dt$ is very small is quite reasonable and, indeed, the Michaelis–Menten formula is widely used in describing enzymatic processes.

But what about Fig. 2.1b? Such a profile is based on a different enzymatic process, for example, when an enzyme E can bound first with one substrate S and then with another substrate S. Furthermore, in such a case, as is well established experimentally, the speed by which the enzyme bounds with the second substrate is much faster. We model such processes as follows:

$$
\begin{aligned}
S + E \quad &\overset{k_1}{\underset{k_{-1}}{\rightleftharpoons}} \quad C_1, && (C_1 = SE) \\
C_1 \quad &\overset{k_2}{\longrightarrow} \quad E + P, && \\
S + C_1 \quad &\overset{k_3}{\underset{k_{-3}}{\rightleftharpoons}} \quad C_2, && (C_2 = SC_1 = S^2 E) \\
C_2 \quad &\overset{k_4}{\longrightarrow} \quad C_1 + P
\end{aligned}
\qquad (2.0.13)
$$

so that

$$\frac{d[P]}{dt} = k_2[C_1] + k_4[C_2].$$

Note that $[E] + [C_1] + [C_2] = const. = e_0$. Assuming the steady state approximations

$$\frac{d[C_1]}{dt} = \frac{d[C_2]}{dt} = 0$$

one can show (see Problem 2.3) that

$$\frac{d[P]}{dt} = \frac{(k_2 K_2 + k_4[S])e_0[S]}{K_1 K_2 + K_2[S] + [S]^2}, \qquad (2.0.14)$$

where

$$K_1 = \frac{k_{-1} + k_2}{k_1}, \quad K_2 = \frac{k_{-3} + k_4}{k_3}.$$

Steps 1 and 3 in Eq. (2.0.13) represent sequential binding of two substrate molecules to the enzyme. We assume that previously enzyme-bound substrate molecule significantly increases the rate of bounding of a second substrate molecule, so that $k_3 \gg k_1$. In the extreme case of $k_1 \to 0$, $k_3 \to \infty$ with $k_1 k_3$ a finite positive constant, we get $K_1 \to \infty$, $K_2 \to 0$, $K_1 K_2 \to K_H > 0$, so that

$$\frac{d[P]}{dt} = \frac{V_{max}[S]^2}{K_H + [S]^2} \tag{2.0.15}$$

where V_{max} and K_H are constants. Formula (2.0.15) is called the **Hill kinetics**; it displays a profile similar to Fig. 2.1b.

Some enzymes can bound with three or more substrates. In this case it is often the case that when enzyme has already bounded with m substrates S, it has a greater affinity to bound with the next substrate S. Under this biological assumption, one can derive the Hill kinetics of order n:

$$\frac{d[P]}{dt} = \frac{V_{max}[S]^n}{K_H + [S]^n}. \tag{2.0.16}$$

Problem 2.4 which will be mentioned in the following suggests how to mathematically prove formula (2.0.16). The profile of $[S] \to [P]$ in Hill kinetics for large n makes a very fast transition from very slow speed to saturated (or limit) speed. The Hill kinetics of order $n \geq 3$ is sometimes used in modeling biochemical reactions within a cell.

PROBLEM 2.1. Consider the chemical reactions

$$A + B \xrightarrow{k} C, \quad B + C \xrightarrow{k} A$$

with $[A] + [C] = 3$ at time $t = 0$. Show that $y = [B]$ satisfies:

$$y(t) = y_0 e^{-3kt}.$$

PROBLEM 2.2. Consider the chemical reactions (2.0.1), (2.0.2) and set $[A] + [C] = m, 2[A] - [B] = n$ at time $t = 0$. Show that the function $y = [B]$ satisfies a differential equation

$$\frac{dy}{dt} = ay^3 + by^2 + cy + d,$$

and compute a, b, c, d in terms of m, n.

PROBLEM 2.3. Derive the relation (2.0.14) under the steady-state approximations $d[C_1]/dt = 0$, $d[C_2]/dt = 0$.

PROBLEM 2.4. Consider the enzymatic process

$$S + E \underset{k_{-1}}{\overset{k_1}{\rightleftharpoons}} C_1, \qquad C_1 \xrightarrow{k_2} E + P,$$

$$S + C_1 \underset{k_{-3}}{\overset{k_3}{\rightleftharpoons}} C_2, \qquad C_2 \xrightarrow{k_4} C_1 + P,$$

$$\vdots$$

$$S + C_{n-1} \underset{k_{-(2n-1)}}{\overset{k_{(2n-1)}}{\rightleftharpoons}} C_n, \qquad C_n \xrightarrow{k_{2n}} C_{n-1} + P.$$

Under the assumptions that $k_1 < k_3 < \ldots < k_{2n-1}$ and $k_1 \to 0$, $k_{2n-1} \to \infty$, and $d[C_j]/dt = 0$ for $1 \le j \le n$, show how to derive the Hill kinetics

$$\frac{d[P]}{dt} = \frac{V_{max}[S]^n}{K_H + [S]^n}.$$

[Hint: $k_{2i-1}[S][C_{i-1}] - (k_{-(2i-1)} + k_{2i})[C_i] = 0$ with $[C_0] = [E]$; hence $[C_j] = \prod_{i=j+1}^{n} K_i [S]^{-(n-j)}[C_n]$, where $K_i = \frac{k_{-(2i-1)}+k_{2i}}{k_{2i-1}}$. Note that $[E] + \sum_{j=1}^{n}[C_j] = e_0, K_1 \to \infty, K_n \to 0$ and assume that the K_j are bounded if $2 \le j \le n-1$ and $K_1 K_2 \cdots K_n \to K_H$.]

The law of mass action and the Michaelis–Menten law can be applied also to any large population of species, for instance, to cells. Consider two populations of cells: tumor cells with number density T and effector cells of the immune system (such as cytotoxic CD8$^+$ T cells) with number density E. When an effector cell senses a tumor cell in its vicinity, it secrets toxic molecules which kill the tumor cells. We can describe this process schematically by

$$E + T \xrightarrow{k_1} C \xrightarrow{k_2} E + P$$

where C denotes the number density of tumor cells "surrounded" by toxic-secreting effector cells and P is the number density of dead tumor cells.

Applying the law of mass action we can write

$$\frac{d[E]}{dt} = -k_1[E][T] + k_2[C],$$

$$\frac{d[C]}{dt} = k_1[E][T] - k_2[C]$$

so that $[E] + [C] = const. = e_0$. Assuming that the killing of the tumor cells is an enzymatic process, we approximately have $d[C]/dt = 0$, so that

$$k_1[E][T] = k_2[C] = k_2(e_0 - [E]),$$

or

$$[E] = \frac{k_2 e_0}{k_2 + k_1[T]}.$$

It follows that
$$\frac{d[P]}{dt} = k_2[C] = \frac{(k_1 k_2 e_0)\,[T]}{k_2 + k_1[T]}$$
which is the Michaelis–Menten law.

2.1. Introduction to MATLAB

It is well known that closed-form or analytic solutions may or may not exist when one tries to solve algebraic equations or differential equations. Even if the closed-form solutions do exist, the derivation may be quite cumbersome. Due to the invention of computers, modern computation techniques and algorithms enable us to seek for solutions in an efficient and robust way. When the closed-form solutions exist, one can perform **symbolic** calculation on computers to obtain the expression of solutions. When the closed-form solutions do not exist, one can look for numerical approaches that provide approximations of the exact solutions. Since numerical solutions are not exact, it is important to understand how accurate numerical solutions are and how robust the numerical approaches are. A numerical approach can be implemented in many different programming languages, such as C, Fortran, Java, MATLAB, Maple, and Mathematica. We will use MATLAB, due to its simplicity and flexibility to code, and provide detail instructions on solving biological models introduced in this book.

MATLAB (MATrix LABoratory) is a high-level numerical computing environment which is developed by MathWorks. MATLAB contains a lot of built-in functions which allow matrix manipulations, algebraic and differential equations solving, data analysis, visualization, etc. We will first describe some basic manipulations for scalars, vectors, and matrices.

To launch MATLAB which is already installed and ready for usage, click on a MATLAB icon and wait for the default user interface to appear. Take the version MATLAB R2012a [17] as an example. The layout of the interface contains several subwindows, including a current directory address on the top to inform you which directory you are currently working on, a current folder on the left which lists the files in the current folder, a command window in the center which allows you to perform calculations, a workspace on the top right to indicate variables you use and their values, and a command history which keeps track of all command lines you type in the command window. If you are a beginner in MATLAB, click on the rightmost item "Help" in the manual bar on top of the directory address. Select "Product Help," double click on "MATLAB," double click on "Getting Started," and then double click on "Quick Start." Here you can learn about the topics such as "Matrices and Arrays," "Functions," and "Plots" to perform some basic calculations and visualizations in MATLAB.

2.1.1. Basic Arithmetic Operations, Creating Variables, and Calling Functions. To perform the calculation, we need to type the commands into the command window where you see the prompt sign ">>."

You can perform the basic arithmetic operations such as addition, subtraction, multiplication, and division by using "+," "−," "*," and "/". For example, you can calculate $2.3 \times 5 + 6 \div 3 - 2$ by entering
>> 2.3*5+6/3-2

Algorithm 1 BasicArithmicOperation.m

---

function [f, g] = BasicArithmicOperation(a,b,c,d,e)
% This function has five inputs a, b, c, d, and e and two outputs f, and g.
% It perform two calculations: f = a*b+c/d-e and g = c-b/e.
f = a*b+c/d-e;
g = c-b/e;
end

After you hit the enter (return) key, the solution ans = 11.5000 will show up and another prompt sign will appear to wait for your next command. Notice that "ans" will be added to the workspace as a variable.

You can also enter numbers in scientific notation, e.g., 1.1×10^{-9}, by entering
>> 1.1*10^-9
The solution ans = 1.1000e-09 will show up in the command window. You can also enter 1.1e-9 or 1.1E-9. Notice that "ans" variable in the workspace is replaced by 1.1000e-09. This implies the current value of "ans" has overwritten the previous value. The default format to represent a number in MATLAB is "format short" which is the scaled fixed-point format with five digits. One can switch to the scaled fixed-point format with 15 digits for double precision and seven digits for single precision by entering "format long."

Sometimes it is more convenient to perform the calculations by using variables or calling a function, especially when the same variables or calculations need to be used multiple times. For example, we can create variables "a," "b," "c," "d," and "e" and assign the values as 2.3, 5, 6, 3, and 2 by entering
>> a = 2.3, b = 5, c = 6, d = 3, e=2
To compute 2.3*5+6/3-2 and assign it as the variable "f," we can then enter
>> f = a*b+c/d-e
The solution f = 11.5000 will show up in the common window.

Since the variables from "a" to "f" have been stored in the workspace, we can perform other calculations, e.g.,
>> g = c-b/e
which will return g = 3.5000.

To create a function (subroutine) to perform both calculations f = a*b+c/d-e and g = c-b/e, click "File," move your mouse to "NEW," and then click on "Function." A new window will pop out and allow you to write a function and save it as a script M-file. Type in exactly the same commands shown in Algorithm 1 and save it as "BasicArithmicOperation.m." In MATLAB script M-file, all characters from the % to the end of the line are treated as a comment and will not be executed. Now you can call this function in the command window by entering

>> [f, g] = BasicArithmicOperation(2.3,5,6,3,2)

MATLAB function	Description
abs(x)	Absolute value of x
sqrt(x)	Square root of x
sin(x)	Sine of x in radians
sind(x)	Sine of x in degrees
asin(x)	Inverse sine of x in radians
asind(x)	Inverse sine of x in degrees
sinh(x)	Hyperbolic sine of x in radians
asinh(x)	Inverse hyperbolic sine of x in radians
cos(x)	Cosine of x in radians
cosd(x)	Cosine of x in degrees
tan(x)	Tan of x in radians
cot(x)	Cotangent of x in radians
sec(x)	Secant of x in radians
csc(x)	Cosecant of x in radians
exp(x)	Exponential of x
log(x)	Natural logarithm of x
log2(x)	Base 2 logarithm of x
log10(x)	Base 10 logarithm of x
floor(x)	Round x toward minus infinity
ceil(x)	Round toward plus infinity
round(x)	Round toward nearest integer
fix(x)	Round toward zero

TABLE 2.1. Some built-in MATLAB functions

MATLAB will return f = 11.5000 and g = 3.5000. If you want to perform this calculation with a = 1, b = 2, c = 3, d = 4, and e = 5, you can simply perform that by entering

>> [f, g] = BasicArithmicOperation(1,2,3,4,5)

MATLAB will return f = -2.2500 and g = 2.6000.

MATLAB contains a lot of built-in functions which can execute in the command window or in the script M-file you would like to generate. In Table 2.1, some commonly used functions are listed. For example, in the command window, you can enter

>> log2(16)

which will return ans = 4.

2.1.2. Vector and Matrix Calculations. Since most of the biological models are formulated as system of equations, we need to know how to perform vector and matrix calculations on MATLAB. To create a row vector with three elements 1, 2.5, and 4, separate the elements with either a comma or a space. Enter either

>> a = [1, 2.5, 4]

or

>> a = [1 2.5 4]

which will return

a =

1.0000 2.5000 4.0000

To create a column vector with the same three elements, separate the elements with semicolons:

>> a = [1; 2.5; 4]

which will return

a =

1.0000

2.5000

4.0000

To create a matrix that has multiple rows, separate the rows with semicolons. For example, to create a 3-by-3 matrix with 1 to 9 integers, enter

>>a = [1 2 3; 4 5 6; 7 8 9]

which will return

a =

1 2 3

4 5 6

7 8 9

MATLAB contains several built-in functions to create fundamental matrices such as ones, zeros, eye, rand, randi, or randn. For example, entering

>>eye(5)

will create a 5-by-5 identity matrix.

To perform basic matrix arithmetic operations such as addition, subtraction, and multiplication, simply use "+," "−," and "*". For example, enter

>> A = [1 2; 3 4]; B = [5 6 ;7 8];

to create two 2-by-2 matrices A and B. If we want to compute C = AB, enter

>> C = A*B

which will return

C =

19 22

43 50

Notice that the matrix product is not commutative. If we compute D = BA, enter

>> D = B*A

which will return
D =
23 34
31 46

To perform element-wise multiplication rather than matrix multiplication, use the .* operator
>> E = A.*B

MATLAB script	Description
A' or ctranspose(A)	Complex conjugate transpose of A
A.' or transpose(A)	Non-conjugate transpose of A
inv(A)	Matrix inverse of the square matrix A
norm(A) or norm(A,2)	2-norm of A
norm(A,1)	1-norm of A
norm(A,Inf)	Infinity norm of A
norm(A,'fro')	Frobenius norm of A
rank(A)	An estimate of the number of linearly independent rows or columns of A
size(A)	Size of A
numel(A)	Total number of elements in a vector or matrix

TABLE 2.2. Some manipulations of a given matrix A

which will return
E =
5 12
21 32

Some other important matrix manipulations are listed in Table 2.2.

MATLAB also provides various ways to solve linear system of equations [19]. X = linsolve(A,B) solves the linear system $AX = B$ using LU factorization with partial pivoting when A is square and QR factorization with column pivoting otherwise. Warning is given if A is ill conditioned for square matrices and rank deficient for rectangular matrices. For example,
>> A = [1 2 ;3 4]; B = [4;10]; X = linsolve(A,B)
which will return
X =
2.0000
1.0000

One can also use backslash ("\") to solve $AX = B$ by entering
>> X = A\B
which will also return
X =
2.0000
1.0000

The backslash computes the solution in a different way. To check out the detail description, enter

>> help \

which will provide the usage of backslash ("\") and the method for numerical implementation.

2.1.3. Symbolic Calculation. Symbolic Math Toolbox [18] provides functions for solving and manipulating symbolic math expressions and performing variable-precision arithmetic. However, it requires a different license from the main MATLAB package. If it is installed on your computer, it allows you to analytically perform differentiation, integration, simplification, transformation, and equation solving.

To declare variables x and y as symbolic objects, use the syms command:

>> syms x y

You can then perform mathematical calculation symbolically. For example, $x + y + 4y$ can be computed by entering

>> x+y+4*y

which will return

ans =

x + 5*y

Symbolic Math Toolbox also enables you to convert numbers to symbolic objects. To create a symbolic number, use the sym command. For example, when we want to compute $\sqrt{2}$ by using double precision, enter

>> a = sqrt(2)

which will return

ans =

1.4142

However, when we want to perform the calculation symbolically, enter

>> a = sqrt(sym(2))

which will return

a = 2^(1/2)

To solve an algebraic equation, one can use the built-in function "solve". For example, to solve $ax^2 + bx + c = 0$, one can enter

>> syms a b c x; solve(a*x^2 + b*x + c)

which will return

ans =

-(b + (b^2 - 4*a*c)^(1/2))/(2*a)
-(b - (b^2 - 4*a*c)^(1/2))/(2*a)

To solve a system of algebraic equations, one can use "," to separate each algebraic equation. For example, to solve

$$\begin{cases} x^2 + xy + y &= 7, \\ x^2 - 4x - y &= -5, \end{cases}$$

and assign the solution to variables Sx and Sy, one can enter

>> syms x y; [Sx,Sy] = solve(x^2 + x*y + y == 7,x^2 - 4*x -y == -5)

which will return

Sx =

2

-i

i

Sy =

1

4 + 4*i

4 - 4*i

where $i = \sqrt{-1}$ is the imaginary unit.

To solve a differential equation [2], one can use the built-in function "dsolve". For example, to solve $\frac{dx(t)}{dt} = ax(t)$ without specified initial condition, enter

>>syms x(t) a; dsolve(diff(x) == a*x)

which will return

ans =

C1*exp(a*t)

where C1 may be replaced by another constant. To solve $\frac{dx(t)}{dt} = ax(t)$ with the initial condition $x(0) = 1$ and assign it to Sx, enter

>> syms x(t) a; Sx = dsolve(diff(x) == a*x,x(0)==1)

which will return

Sx =

exp(a*t)

To solve a system of differential equations

$$\begin{cases} \frac{df(t)}{dt} &= f(t) + 2g(t), \\ \frac{dg(t)}{dt} &= -f(t) + 2g(t), \end{cases} \tag{2.1.1}$$

and assign the solution to S, enter

>>syms f(t) g(t); S = dsolve(diff(f) == f + 2*g, diff(g) == -f + 2*g)

which will return

S =

g: [1x1 sym]

f: [1x1 sym]

To return the values of $g(t)$, enter the command:

>>g(t) = S.g

which will return

g(t) = C2*exp((3*t)/2)*cos((7^(1/2)*t)/2) - C3*exp((3*t)/2)*
     sin((7^(1/2)*t)/2)

where C2 and C3 may be replaced by other constants.

To solve (2.1.1) with initial conditions $f(0) = 1$ and $g(0) = 2$, enter

>>syms f(t) g(t); S = dsolve(diff(f) == f + 2*g, diff(g) == -f + 2*g,
f(0)==1, g(0)==2);

>>g(t) = S.g

g(t) =

2*cos((7^(1/2)*t)/2)*exp(t)^(3/2)

PROBLEM. 2.5. Write scripts in MATLAB to solve Problems 2.1 and 2.2 symbolically.

To exit MATLAB, you can either click on the red button at the top left corner or simply type quit in the command window.

CHAPTER 3

Ordinary Differential Equations

In this chapter we review some basic facts about systems of ordinary differential equations of the form

$$\frac{dx_i}{dt} = f_i(x_1, \ldots, x_n) \quad (1 \le i \le n),$$

or, in vector form,

$$\frac{d\mathbf{x}}{dt} = \mathbf{f}(\mathbf{x}) \tag{3.0.1}$$

where $\mathbf{x} = (x_1, \ldots, x_n)$, $\mathbf{f} = (f_1, \ldots, f_n)$. The following theorem is well known [5, Chapter 1].

THEOREM 3.1. *If the functions $f_i(\mathbf{x})$ and their first derivatives $\partial f_i(\mathbf{x})/\partial x_j$ are continuous for $-\infty < x_k < \infty$ ($k = 1, \ldots, n$), then for any initial condition*

$$\mathbf{x}(0) = \mathbf{x}_0 \tag{3.0.2}$$

there exists a unique solution of (3.0.1), (3.0.2) for a small time interval $-\delta < t < \delta$.

The solution can be continued to any interval $-r_1 < t < r_2$ as long as $\mathbf{x}(t)$ remains bounded [5, Chap. 1]. In particular:

THEOREM 3.2.

    (i) *If $\mathbf{f}(\mathbf{x})$ is bounded linearly, that is,*

$$|\mathbf{f}(\mathbf{x})| \le c_1|\mathbf{x}| + c_2 \tag{3.0.3}$$

        for some positive constants c_1 and c_2, then the solution can uniquely be extended to all $-\infty < t < \infty$.

    (ii) *If $\mathbf{f}(\mathbf{x})$ is not bounded linearly, but*

$$\mathbf{x} \cdot \mathbf{f}(\mathbf{x}) \le c_1|\mathbf{x}|^2 + c_2 \tag{3.0.4}$$

        for some positive constants c_1 and c_2, then the solution $\mathbf{x}(t)$ can uniquely be extended to $-\delta < t < \infty$.

In the last theorem, the following notations were used:

$$|\mathbf{z}| = \left(\sum_{i=1}^{n} z_i^2 \right)^{\frac{1}{2}} \quad \text{for} \quad \mathbf{z} = (z_1, \ldots, z_n), \quad \text{and} \quad \mathbf{x} \cdot \mathbf{z} = \sum_{i=1}^{n} x_i z_i.$$

© Springer International Publishing Switzerland 2014
A. Friedman, C.-Y. Kao, *Mathematical Modeling of Biological Processes*, Lecture Notes on Mathematical Modelling in the Life Sciences, DOI 10.1007/978-3-319-08314-8_3

PROBLEM 3.1. Prove that under the assumption (3.0.3 [or (3.0.4)]) the solution $x(t)$ remains bounded in any time interval $-T < t < T$ $(0 \leq t < T)$.

Solutions of (3.0.1) are called **trajectories**, and the space of all possible points $\mathbf{x} = (x_1, \ldots, x_n)$ of trajectories of (3.0.1) is called the **phase space**. A trajectory that exists for all $-\infty < t < \infty$ is called an **orbit**.

A point $\mathbf{x}_0$ such that $\mathbf{f}(\mathbf{x}_0) = 0$ is called a **steady point** (**steady state**, an **equilibrium point**, or a **fixed point**) of the ODE (3.0.1). The corresponding solution of (3.0.1), (3.0.2) is $\mathbf{x}(t) \equiv \mathbf{x}_0$. A steady-state $\mathbf{x}_0$ is **stable** if for any small δ_1 there exists a δ_2 such that if $|\mathbf{x}(0) - \mathbf{x}_0| < \delta_2$, then the solution $\mathbf{x}(t)$ exists all $t > 0$ and $|\mathbf{x}(t) - \mathbf{x}_0| < \delta_1$ for all $t > 0$. A stable steady-state $\mathbf{x}_0$ is **asymptotically stable** if any solution $\mathbf{x}(t)$ with $\mathbf{x}(0)$ near $\mathbf{x}_0$ converges to $\mathbf{x}_0$ as $t \to \infty$. The steady state $\mathbf{x}_0$ is **unstable** if it is not stable.

EXAMPLE. The equation

$$\frac{dx}{dt} = f(x) = x - x^3$$

has the steady states $x_0 = 0, \pm 1$. Note that $dx/dt > 0$ if $x < -1$ and $dx/dt < 0$ if $-1 < x < 0$. Hence $x(t) \to -1$ if $t \to \infty$ provided $x(0)$ is near -1, so that $x_0 = -1$ is asymptotically stable. A similar reasoning leads to the conclusion that $x_0 = +1$ is asymptotically stable and $x_0 = 0$ is unstable.

If $\mathbf{x}_0$ is a steady point of Eq. (3.0.1), one can write

$$\mathbf{f}(\mathbf{x}) = A(\mathbf{x} - \mathbf{x}_0) + (\mathbf{x} - \mathbf{x}_0)o(|\mathbf{x} - \mathbf{x}_0|) \tag{3.0.5}$$

where $o(|\xi|) \to 0$ if $|\xi| \to 0$ and A is the Jacobian matrix with elements

$$a_{ij} = \frac{\partial f_i}{\partial x_j}(\mathbf{x}_0).$$

Consider the linear system

$$\frac{d\mathbf{y}}{dt} = A\mathbf{y} \tag{3.0.6}$$

where $\mathbf{y} = (y_1, \ldots, y_n)$. A function $\mathbf{y}(t) = \mathbf{y}_0 e^{\lambda t}$ is a solution of (3.0.6) if and only if

$$A\mathbf{y}_0 = \lambda \mathbf{y}_0. \tag{3.0.7}$$

This equation has a solution with $\mathbf{y}_0 \neq 0$ if and only if λ satisfies the **characteristic equation**

$$\det(\lambda I - A) \equiv \lambda^n + \alpha_1 \lambda^{n-1} + \ldots + \alpha_{n-1}\lambda + \alpha_n = 0. \tag{3.0.8}$$

Such a λ is called an **eigenvalue** and the corresponding vector $\mathbf{y}_0$ is called an **eigenvector**.

If all the eigenvalues of A are different, then the general solution of (3.0.7) is

$$\mathbf{y}(t) = \sum_{j=1}^{n} c_j \mathbf{y}_{0,j} e^{\lambda_j t} \tag{3.0.9}$$

where the c_j are arbitrary constants and $\mathbf{y}_{0,j}$ is an eigenvector corresponding to the eigenvalue λ_j. If some of the eigenvalues coincide, say $\lambda_1 = \lambda_2 = \ldots = \lambda_k$, then one needs to replace c_j by $c_j t^{j-1}$ $(j = 1, \ldots, k)$ to obtain the general solution.

It follows that if all the eigenvalues have negative real parts, then any solution $\mathbf{y}(t)$ of (3.0.6) converges to $\mathbf{y} = 0$ as $t \to \infty$, so that $\mathbf{y} = 0$ is an asymptotically stable steady state of the linear ODE system (3.0.6). If one of the eigenvalues has positive real part, then $\mathbf{y} = 0$ is unstable: indeed, a solution $\mathbf{y}_0 e^{\lambda t}$ with $\mathrm{Re}(\lambda) > 0$ is arbitrarily close to $\mathbf{y} = 0$ if $t \to -\infty$ but $|\mathbf{y}_0 e^{\lambda t}| \to \infty$ if $t \to \infty$.

For the general case of Eq. (3.0.1) the following theorem holds [5, Chap. 13]:

THEOREM 3.3. *Let $\mathbf{x}_0$ be a steady point of (3.0.1), so that (3.0.5) holds. If all the eigenvalues of the Jacobian matrix A have negative real parts, then $\mathbf{x}_0$ is asymptotically stable, and $|\mathbf{x}(t) - \mathbf{x}_0| \le const.e^{-\mu t}$ $(\mu > 0)$ for all $t > 0$ if $|\mathbf{x}(0) - \mathbf{x}_0|$ is sufficiently small. If at least one of the eigenvalues has a positive real part, then $\mathbf{x}_0$ is unstable; indeed, there exists a positive constant δ and $\mathbf{x}(0)$ with arbitrarily small $|\mathbf{x}(0) - \mathbf{x}_0|$, such that $|\mathbf{x}(t) - \mathbf{x}_0|$ cannot remain in $\{|\mathbf{x}(0) - \mathbf{x}_0| < \delta\}$ for all $t > 0$.*

For the case of a linear system of two equations,

$$
\begin{aligned}
\frac{dx_1}{dt} &= a_{11}x_1 + a_{12}x_2, \\
\frac{dx_2}{dt} &= a_{21}x_1 + a_{22}x_2,
\end{aligned}
\tag{3.0.10}
$$

with the 2×2 matrix

$$
A = \begin{pmatrix} a_{11} & a_{12} \\ a_{21} & a_{22} \end{pmatrix},
$$

the characteristic equation is

$$
\lambda^2 - (\mathrm{trace}A)\,\lambda + \det A = 0
\tag{3.0.11}
$$

where $\mathrm{trace}A = a_{11} + a_{22}$, $\det A = a_{11}a_{22} - a_{12}a_{21}$. Then the two eigenvalues have negative real parts if and only if

$$
\mathrm{trace}A < 0 \text{ and } \det A > 0.
$$

The behavior of the trajectories of (3.0.10) for general eigenvalues λ_1, λ_2 of the characteristic equation (3.0.11), as time t increases, can be described by a **phase portrait**, shown in Fig. 3.1.

In the general case $n \ge 3$, there is a simple criterion to determine whether all the eigenvalues of a polynomial

$$
\alpha_0 \lambda^n + \alpha_1 \lambda^{n-1} + \ldots + \alpha_{n-1}\lambda + \alpha_n \quad (\alpha_0 = 1)
\tag{3.0.12}
$$

have negative real parts. This is stated in terms of the matrices

$$A_1 = (\alpha_1), \quad A_2 = \begin{pmatrix} \alpha_1 & \alpha_0 \\ \alpha_3 & \alpha_2 \end{pmatrix}, \quad A_3 = \begin{pmatrix} \alpha_1 & \alpha_0 & 0 \\ \alpha_3 & \alpha_2 & \alpha_1 \\ \alpha_5 & \alpha_4 & \alpha_3 \end{pmatrix},$$

$$A_j = \begin{pmatrix} \alpha_1 & \alpha_0 & 0 & \cdots & 0 \\ \alpha_3 & \alpha_2 & \alpha_1 & \cdots & 0 \\ \alpha_5 & \alpha_4 & \alpha_3 & \cdots & \\ \vdots & & \vdots & & \vdots \\ \alpha_{2j-1} & \alpha_{2j-2} & \alpha_{2j-3} & \cdots & \alpha_j \end{pmatrix} \quad (4 \le j \le n)$$

where the (l, m) term in the matrix A_j is defined as follows:

$$\begin{array}{ll} \alpha_{2l-m} & \text{if} \qquad\qquad 0 < 2l - m \le n, \\ \alpha_0 & \text{if} \qquad\qquad 2l - m = 0, \\ 0 & \text{if} \quad 2l - m < 0 \quad \text{or} \quad 2l - m > n. \end{array}$$

a   Unstable Node
($\lambda_1 > 0$, $\lambda_2 > 0$)

b   Stable Node
($\lambda_1 < 0$, $\lambda_2 < 0$)

c   Saddle Point
($\lambda_1 > 0$, $\lambda_2 < 0$)

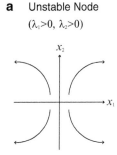
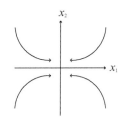
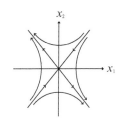

d   Unstable Spiral
($\lambda_1 = \alpha + i\beta$, $\alpha > 0$)

e   Stable Spiral
($\lambda_1 = \alpha + i\beta$, $\alpha < 0$)

f   Center
($\lambda_1 = i\beta$)

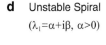
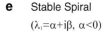
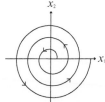
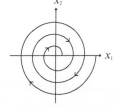
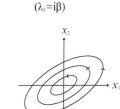

FIGURE 3.1. Phase portrait. (**a**) Unstable node ($\lambda_1 > 0$, $\lambda_2 > 0$). (**b**) Stable node ($\lambda_1 < 0, \lambda_2 < 0$). (**c**) Saddle point ($\lambda_1 > 0, \lambda_2 < 0$). (**d**) Unstable spiral ($\lambda_1 = \alpha + i\beta, \alpha > 0$). (**e**) Stable spiral ($\lambda_1 = \alpha + i\beta, \alpha < 0$). (**f**) Center ($\lambda_1 = i\beta$)

THEOREM 3.4. *(Routh–Hurwitz). All the zeros of the polynomial (3.0.12) have negative real parts if and only if*

$$\det(A_j) > 0 \quad \text{for} \quad j=1,2,\ldots,n. \tag{3.0.13}$$

In case $n = 3$ the Routh–Hurwitz conditions, with $\alpha_0 = 1$, are

$$\alpha_1 > 0, \alpha_3 > 0, \alpha_1\alpha_2 > \alpha_3, \tag{3.0.14}$$

and, in case $n = 4$, they are

$$\alpha_1 > 0, \alpha_3 > 0, \alpha_4 > 0, \alpha_1\alpha_2\alpha_3 > \alpha_3^2 + \alpha_1^2\alpha_4. \tag{3.0.15}$$

A proof of Theorem 3.4 is given in [12].

PROBLEM 3.2. Prove that the Routh–Hurwitz conditions (3.0.13) with $\alpha_0 = 1$ reduce to (3.0.14) in the case $n = 3$ and to (3.0.15) in the case $n = 4$.

PROBLEM 3.3. Find the general solution of

$$\frac{dx_1}{dt} = -2x_1 + 7x_2,$$
$$\frac{dx_2}{dt} = 2x_1 + 3x_2.$$

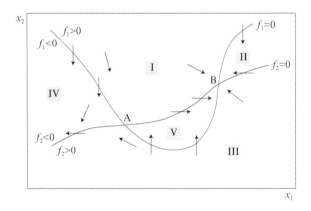

FIGURE 3.2. Schematic illustration of nullclines (*curves where $dx_1/dt = f_1 = 0$ and $dx_2/dt = f_2 = 0$*) and directions of the vector field (*arrows*)

PROBLEM 3.4. Find the general solution of

$$\frac{dx_1}{dt} = x_1 - 2x_2,$$
$$\frac{dx_2}{dt} = 2x_1 + x_2.$$

PROBLEM 3.5. Determine whether the steady states of the following systems are asymptotically stable:

(a) $\frac{dx}{dt} = y - rx + x^2$, $\frac{dy}{dt} = x - y - 1$, $(r > 0)$.

(b) $\frac{dx}{dt} = x + y - z$, $\frac{dy}{dt} = x - y + z^2$, $\frac{dz}{dt} = 2x - z^2 + 1$.

Consider a system of two equations:

$$\begin{array}{rcl} \frac{dx_1}{dt} &=& f_1(x_1, x_2) \\ \frac{dx_2}{dt} &=& f_2(x_1, x_2) \end{array} \qquad (3.0.16)$$

The curve $f_i(x_1, x_2) = 0$ is called the x_i- **nullcline**. The steady points of the system (3.0.13) are the points where the two nullclines intersect.

Figure 3.2 depicts a situation where the two nullclines have two intersection points, A and B. The arrows indicate the direction of the vector field (f_1, f_2). It is seen that the steady state B is stable and A is unstable. The **phase diagram**, i.e., the geometric description of trajectories in the phase space, also shows that even if the initial value $\mathbf{x}(0)$ is not near B, the solution may nevertheless converge to B as $t \to \infty$. For example, if $\mathbf{x}(0)$ belongs to the region II or V, then $\mathbf{x}(t) \to B$ as $t \to \infty$.

In Fig. 3.3 the steady states A and C are stable, and the state B is unstable. A system with two stable steady states is called **bistable**.

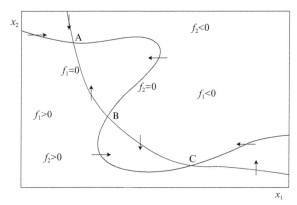

FIGURE 3.3. Nullclines with three intersections corresponding to two stable steady states (A and C) and one unstable steady state (B). Directions of vector fields are shown by *arrows*

3.1. Cancer Model

In Chap. 2 we modeled a process by which effector cells (E) of the immune system kill tumor cells (T) as they come into contact with them:

$$E + T \xrightarrow{k_1} C \xrightarrow{k_2} E + P$$

where C is the population of tumor cells that are being killed and P is the population of dead tumor cells. We deduced the Michaelis–Menten formula (with $k_{-1} = 0$)

$$\frac{dP}{dt} = \frac{(k_1 k_2 e_0)\, T}{k_2 + k_1 T},$$

where e_0 is the total population of effector cells; for simplicity we dropped the square brackets around T and P. In the absence of effector cells, the tumor cells will grow at some rate λ, but constrained by space and resources, their proliferation is modeled by the **logistic growth**

$$\frac{dT}{dt} = \lambda T \left(1 - \frac{T}{T_0} \right),$$

where T_0 is the capacity of the tissue for sustaining the tumor population. Taking the immune response into account, the complete model for tumor growth is then

$$\frac{dT}{dt} = \lambda T \left(1 - \frac{T}{T_0} \right) - \frac{(k_1 k_2 e_0)\, T}{k_2 + k_1 T} \equiv f(T). \qquad (3.1.1)$$

Note that $T(t)$ will remain less than T_0 if $T(0) < T_0$. The effective killing rate of the tumor cells by the total population of the effector cells is $k_1 e_0$, while the tumor growth rate is λ. We consider two cases:

(i)
$$k_1 e_0 > \lambda \quad \text{(effective killing exceeds growth)}$$

(ii)
$$k_1 e_0 < \lambda \quad (\text{ growth exceeds effective killing})$$

In the first case we expect that cancer will not develop, while in the second case we expect that tumor cells may grow and cancer will develop. Here we use the words "tumor" and "cancer" interchangeably; more commonly, cancer usually refers to tumor that has metastasized.

Case (i). Note that $f(0) = 0$ and

$$f'(0) = \lambda - \frac{k_1 k_2 e_0}{k_2} = \lambda - k_1 e_0 < 0$$

in case (i). Hence $T = 0$ is asymptotically stable. Thus, any initially small tumor will decrease to zero as t increases to ∞.

Case (ii). In this case $f'(0) > 0$ so that $T = 0$ is not stable, and small tumors will not shrink to zero. In order to determine the behavior of $T(t)$ as $t \to \infty$, we rewrite (3.1.1) in the form

$$\frac{dT}{dt} = \frac{\lambda T}{k_2 + k_1 T} h(T),$$

where

$$h(T) = -\frac{k_1 T^2}{T_0} + \left(k_1 - \frac{k_2}{T_0} \right) T + k_2 \left(1 - \frac{k_1 e_0}{\lambda} \right).$$

The function $h(T)$ is a parabola, with

$$h(-\infty) = h(+\infty) = -\infty,$$

and

$$h(0) = k_2 \left(1 - \frac{k_1 e_0}{\lambda}\right) > 0, \quad h(T_0) = -\frac{k_1 k_2 e_0}{\lambda} < 0.$$

Hence

$$\begin{aligned} h(T) &> 0 \quad \text{if} \quad 0 < T < T_1, \\ h(T) &< 0 \quad \text{if} \quad T > T_1, \end{aligned}$$

and therefore also

$$\begin{aligned} f(T) &> 0 \quad \text{if} \quad 0 < T < T_1, \\ f(T) &< 0 \quad \text{if} \quad T > T_1, \end{aligned}$$

where T_1 is the positive solution of $h(T) = 0$, $T_1 < T_0$, and $h'(T_1) < 0$ It follows that if $T(0) < T_1$, then $T(t) \uparrow T_1$ as $t \uparrow \infty$, and if $T_1 < T(0) < T_0$, then $T(t) \downarrow T_1$ as $t \uparrow \infty$.

3.2. Root-Finding Algorithms

To find steady states of a system of ordinary differential equations of the form (3.0.1), we introduce methods to solve,

$$f_i(x_1, \ldots, x_n) = 0, \quad i = 1, \ldots, n. \tag{3.2.1}$$

Let us start with one scalar equation in a single variable,

$$f(x) = 0. \tag{3.2.2}$$

The algorithms that find the value x (a root) such that $f(x) = 0$ are called **root-finding** algorithms [7, Chap. 6] [2, Chap. 2]. Two of the best known root-finding algorithms are the **bisection method** and the **Newton's method**, named after the eminent eighteenth-century mathematician and scientist Isaac Newton. The bisection method is a "gradient-free" approach and usually takes longer to converge, but it is more robust. The Newton's method uses gradient (slope in one dimension) information and is more efficient; however, it may fail when the initial estimate is too far away from the root.

3.2.1. Bisection Method. The idea of the bisection method comes from the **intermediate value theorem** which states the continuous function f must have at least one root in the interval (a, b) if $f(a)$ and $f(b)$ have opposite signs. The method repeatedly bisects an interval and then selects, for further processing, a subinterval in which a root must lie. Suppose that we have two initial points $a_0 = a$ and $b_0 = b$ such that $f(a)f(b) < 0$. The method divides the interval into two by computing the midpoint $c = \frac{a+b}{2}$ of the interval. If c is a root, then the algorithm terminates; otherwise, the algorithm checks whether $f(a)f(c)$ or $f(c)f(b)$ is negative. If, for example,

$f(a)f(c) < 0$, the root must lie in the interval (a, c) and the method sets a as a_1 and c as b_1. Repeating this process, we can construct a sequence of intervals $[a_n, b_n]$ such that

$$|b_n - a_n| = \frac{|b_0 - a_0|}{2^n}.$$

Since the root must lie in these subintervals, the best estimate for the location of the root is the midpoint of the smallest subinterval found. In that case, the absolute error after n steps is at most

$$\frac{|b - a|}{2^{n+1}}. \tag{3.2.3}$$

If either endpoint of the smallest interval is used, then the maximum absolute error is

$$\frac{|b - a|}{2^n}. \tag{3.2.4}$$

If we use (3.2.4) to determine the number of steps such that the error is smaller than a given tolerance ϵ, the number of iterations needs to satisfy

$$n > \log_2 \frac{|b - a|}{\epsilon}.$$

3.2.2. Newton's Method. Instead of using only the value of the function f, Newton's method uses also the derivative of the function. Given the initial guess x_0, Newton's method generates a sequence of approximations of the root by

$$x_{n+1} = x_n - \frac{f(x_n)}{f'(x_n)} \tag{3.2.5}$$

until a sufficiently accurate value is reached. This idea comes from the linear approximation near the root,

$$f(x_{n+1}) \approx f(x_n) + (x_{n+1} - x_n)f'(x_n).$$

If the function f is continuously differentiable and its derivative does not vanish at the root α and if f has a second derivative in some interval containing α, then the convergence is quadratic. To prove this we use the Taylor's expansion near α,

$$0 = f(\alpha) = f(x_n) + f'(x_n)(\alpha - x_n) + R_1$$

where

$$R_1 = \frac{1}{2}f''(\xi_n)(\alpha - x_n)^2$$

and ξ_n lies between x_n and α. Thus

$$\alpha = x_n - \frac{f(x_n)}{f'(x_n)} - \frac{f''(\xi_n)}{2f'(x_n)}(\alpha - x_n)^2. \tag{3.2.6}$$

Setting $e_n = \alpha - x_n$, and subtracting (3.2.5) from (3.2.6), we have

$$e_{n+1} = -\frac{f''(\xi_n)}{2f'(x_n)}e_n^2.$$

Taking absolute value of both sides gives

$$|e_{n+1}| = \frac{|f''(\xi_n)|}{2|f'(x_n)|}e_n^2.$$

Set

$$M = \sup_{x \in I} \frac{1}{2}\left|\frac{f''(\xi_n)}{f'(x_n)}\right|, \quad I = [\alpha - r, \alpha + r] \text{ for some } r > 0.$$

The necessary condition of convergence for the initial point x_0 is $M|e_n| < 1$. Thus the rate of convergence is quadratic if $f'(x) \neq 0$ for $x \in I$, $f''(x)$ is bounded for $x \in I$, and x_0 sufficiently close to the root α, so that $|x_0 - \alpha| < r$. This requirement does not explicitly tell us how to choose x_0 since we do not know the root α before the computation.

Newton's method can be easily extended to solve the general nonlinear equations (3.2.1). Instead of dividing in (3.2.5) by $f'(x_n)$, one has to left multiply by the inverse of the $n \times n$ Jacobian matrix $F'(X_n)$, i.e.,

$$X_{n+1} = X_n - [F'(X_n)]^{-1} F(X_n). \tag{3.2.7}$$

For numerical purposes it is more common to rewrite (3.2.7) in the form

$$F'(X_n)(X_{n+1} - X_n) = F(X_n).$$

First the linear system

$$F'(X_n)\tilde{X} = F(X_n)$$

is solved for $\tilde{X}$ and then the approximation, in the next step, is obtained by

$$X_{n+1} = \tilde{X} + X_n.$$

3.2.3. Built-In MATLAB Functions for Root Finding. In Sect. 2.1, we have introduced a symbolic way in MATLAB to solve a system of nonlinear equation (3.2.1) by using the built-in function "solve." There are several other functions which allow one to perform the same task. For example, the built-in function "roots(C)" computes the roots of the polynomial whose coefficients are the elements of the vector C. If C has $N + 1$ components, the polynomial is $C(1)x^N + \ldots + C(N)x + C(N+1)$:

>> x=roots([1 -10 35 -50 24])

which returns

x =

4.0000

3.0000

2.0000

1.0000

Another built-in function "fzero" [3] performs a single-variable nonlinear root finding:

$x =$ fzero(fun,x_0)

Iteration	Func-count	f(x)	Norm of step	First-order optimality	Trust-region radius
0	3	47071.2		2.29e+04	1
1	6	12003.4	1	5.75e+03	1
2	9	3147.02	1	1.47e+03	1
3	12	854.452	1	388	1
4	15	239.527	1	107	1
5	18	67.0412	1	30.8	1
6	21	16.7042	1	9.05	1
7	24	2.42788	1	2.26	1
8	27	0.032658	0.759511	0.206	2.5
9	30	7.03149e-06	0.111927	0.00294	2.5
10	33	3.29525e-13	0.00169132	6.36e-07	2.5

```
Equation solved.

fsolve completed because the vector of function values is near zero
as measured by the default value of the function tolerance, and
the problem appears regular as measured by the gradient.
```

TABLE 3.1. Output for fsolve from MATLAB

tries to find a zero of the input function "fun" near x_0, if x_0 is a scalar and fun is a function handle. For example,

$>> x =$ fzero(@cos,1)

$x =$1.5708.

The function "fsolve" solves systems of nonlinear equations of several variables:

$x =$ fsolve(fun,x_0)

starts at x_0 and tries to solve the equations described in fun. For example, solve

$$2x_1 - x_2 = e^{-x_1},$$
$$-x_1 + 2x_2 = e^{-x_2},$$

with the initial condition $[x_1, x_2] = [-5, -5]$. First, write a file that computes F, the values of the equations at x:

function F = myfun(x)

F = [2*x(1) - x(2) - exp(-x(1)); -x(1) + 2*x(2) - exp(-x(2))];

Save this function file as myfun.m somewhere on your MATLAB path. Next, set up the initial point and options and call fsolve:

x0 = [-5; -5]; % Make a starting guess of the solution

options=optimset('Display','iter'); % Option to display output

[x,fval] = fsolve(@myfun,x0,options) % Call the solver

After several iterations, fsolve finds an answer as shown in Table 3.1. The $f(x)$ in Table 3.1 is the sum of squares of function values. For example, when the initial guess is x0 = $[-5;-5]$, F $\approx [-153.4132; -153.4132]$, and $f(x) = (-153.4132)^2 + (-153.4132)^2 \approx 47071.2$.

Indeed fsolve is completed because the vector of function values is near zero as measured by the default value of the function tolerance, and the problem appears regular as measured by the gradient. The solution x and fval are

x =

0.5671

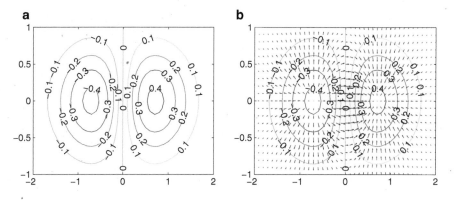

FIGURE 3.4. **(a)** Contour plot **(b)** Vector field plot of $z = xe^{(-x^2 - 2y^2)}$

0.5671
fval =
1.0e-006 *
-0.4059
-0.4059.

Another two MATLAB functions which are useful to study the phase portraits are contour and quiver. For example, the contour plot (shown in Fig. 3.4a) of the function

$$z = xe^{(-x^2 - 2y^2)}$$

over the range $-2 \le x \le 2, -1 \le y \le 1$ with the increment 0.1 in both x- and y-direction and can be done by

>>[X,Y] = meshgrid(-2:0.1:2,-1:0.1:1); % generate the mesh
>>Z = X.*exp(-X.^2-2*Y.^2); % generate the function value
>>[C,h] = contour(X,Y,Z,[-1:0.1:1]); clabel(C,h) % contour plot over the

range -1 to 1 with the increment 0.1.
Now we can add the vector field plot by using quiver

>>[DX,DY] = gradient(Z); % calculate gradient of the function Z
>>hold on; quiver(X,Y,DX,DY) % plot the vector field

From the vector field shown in Fig. 3.4b, we can easily see the stability properties of a steady state.

PROBLEM 3.6. Write algorithms to implement both Bisection method and Newton's method to solve

$$\begin{aligned} x_1^3 + x_2 &= 1, \\ x_2^3 - x_1 &= -1. \end{aligned}$$

Indicate your initial condition and how many steps it requires to reach the tolerance of error to be within 10^{-6}.

PROBLEM 3.7. Choose your own f_1 and f_2 to generate similar phase portraits in Figs. 3.2 and 3.3. Plot the nullclines and indicate the directional field.

CHAPTER 4

Epidemiology of Infectious Diseases

Epidemiology is the study of patterns, causes, and effects of health and disease conditions in a population. It provides critical support for public health by identifying risk factors for disease and targets for preventive medicine.Epidemiology has helped develop methodology used in clinical research and public health studies. Major areas of epidemiological study include disease etiology, disease break, disease surveillance, and comparison of treatment effects such as in clinical trials.

Epidemiologists gather data and apply a broad range of biomedical and psychosocial theories to generate theory, test hypotheses, and make educated, informed assertions as to which relationships are causal and in which way. For example, many epidemiological studies are aimed at revealing unbiased relationships between exposure to smoking, biological agents, stress, or chemicals to mortality and morbidity. In the identification of causal relationship between these exposures and outcome epidemiologists use statistical and mathematical tools.

In this chapter we focus on epidemiology of infectious diseases. The adjectives **epidemic** and **endemic** are used to distinguish between a disease spread by an infective agent (epidemic) and a disease which resides in a population (endemic). For example, there are occasional spreads of the **cholera** epidemic in some countries, while **malaria** is endemic in southern Africa. In order to understand which epidemic will die out and which will become endemic, we need to determine several parameters associated with the disease:

C = the number of contacts an infectious person makes per unit time;
P = the probability of transmission per contact with an infectious person;
D = the duration that an infected person is infectious to others.

The number of expected secondary infections resulting from a single infection case is, roughly,

$$R_0 = C \cdot P \cdot D.$$

We conclude that if $R_0 < 1$, then the infection will die out, whereas if $R_0 > 1$, then the disease will spread in the population, probably becoming endemic.

© Springer International Publishing Switzerland 2014
A. Friedman, C.-Y. Kao, *Mathematical Modeling of Biological Processes*, Lecture Notes on Mathematical Modelling in the Life Sciences, DOI 10.1007/978-3-319-08314-8_4

Data collected from many sources provide estimate for R_0. For example, for measles, R_0 is in the range of 12–18, and for mumps, R_0 is in the range of 4–7.

The quantities C, P, D may be calculated from statistical data, but they do not give us insight into the actual evolution, or spread, of the disease. In order to gain insight into the spread of a disease caused by an infective agent, we need to build a mechanistic model. Such a mathematical model, described by differential equations, can project how an infectious disease will progress and what will be the outcome with or without intervention. Mathematical models make some assumptions and use parameters such as C, P, D above in order to calculate the effect of possible interventions, like mass vaccination programs.

In what follows we shall develop several generic mechanistic models for infectious diseases. We note that R_0 as defined above is rarely observed in the field. So instead we shall calculate R_0 from the mathematical model; this is then a different number from the one defined above, and the better the model's simulations fit with data, the more useful R_0 will be.

We begin with a simple model of a disease in a population of size N. We divide the population into three classes: susceptible S, infected I, and recovered R. Let

β = infection rate,
μ = death rate, the same for all individuals,
ν = recovery rate,
γ = rate by which recovered individuals have lost
    their immunity and became susceptible to the disease.

Then we have the following diagram:

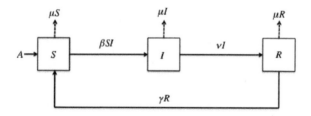

where A is the growth of susceptible. If all newborns are healthy, then, not only S and R, but also I contribute to the growth term A.

Base on the above diagram we set up the following equations:

$$\frac{dS}{dt} = A - \beta SI + \gamma R - \mu S,$$
$$\frac{dI}{dt} = \beta SI - \nu I - \mu I, \qquad (4.0.1)$$
$$\frac{dR}{dt} = \nu I - \gamma R - \mu R.$$

We view each of the populations S, I, R, N as representing a number of individuals (or a number density, i.e., the number of individuals per unit area). The dimension of γ, μ, ν is $1/\text{time}$, the dimension of β is $1/(\text{individual} \cdot \text{time})$, and the dimension of A is $\text{individual}/\text{time}$.

To examine more carefully the meaning of A, we introduce a differential equation for $N(t)$, which is obtained by adding all the equations in (4.0.1),

$$\frac{dN}{dt} = A - \mu N.$$

Given initial population N_0, we find that

$$N(t) = N_0 e^{-\mu t} + \frac{A}{\mu} \left(1 - e^{-\mu t}\right).$$

Hence $N(t) \to A/\mu$ as $t \to \infty$. Thus A is equal to the asymptotic density of the population (as $t \to \infty$) divided by the death rate. The system (4.0.1) is called the **SIR model**.

The SIR model has an equilibrium point which is disease free, namely,

$$(S_0, I_0, R_0) = \left(\frac{A}{\mu}, 0, 0\right);$$

it is called the **disease-free equilibrium (DFE)**. The Jacobian matrix at the DFE is

$$\begin{pmatrix} -\mu & -\beta\frac{A}{\mu} & \gamma \\ 0 & \beta\frac{A}{\mu} - (\nu + \mu) & 0 \\ 0 & \nu & -\gamma - \mu \end{pmatrix}.$$

The characteristic polynomial is

$$(\mu + \lambda) \left(\beta\frac{A}{\mu} - (\nu + \mu) - \lambda\right) (\gamma + \mu + \lambda).$$

Hence the DFE is asymptotically stable if

$$\beta\frac{A}{\mu} < \nu + \mu. \tag{4.0.2}$$

When (4.0.2) holds, any new small infection will die out with time.

On the other hand if

$$\beta\frac{A}{\mu} > \nu + \mu, \tag{4.0.3}$$

the DFE is unstable; there are arbitrarily small infections that will not disappear in the population. Furthermore, there are equilibrium points $(\bar{S}, \bar{I}, \bar{R})$ with $\bar{I} > 0$, namely,

$$\begin{aligned} \beta\bar{S} &= \nu + \mu, \\ \frac{\beta}{\mu}\left((\nu + \mu) - \frac{\gamma\nu}{\gamma + \mu}\right)\bar{I} &= \left(\beta\frac{A}{\mu} - (\nu + \mu)\right), \\ \bar{R} &= \frac{\nu}{(\gamma + \mu)}\bar{I}; \end{aligned} \tag{4.0.4}$$

notice that $\bar{I} > 0$ since

$$\frac{\gamma\nu}{\gamma + \mu} < \nu + \mu.$$

An important concept in epidemiology is the **basic reproduction number**:

In a healthy population we introduce one infection and compute the expected infection among the susceptibles caused by this single infection during its lifetime. We call it the **expected secondary infection**, or **basic reproduction number**, and denote it by R_0. Then intuitively it is clear that DFE is stable if $R_0 < 1$ (the secondary infection will be smaller than the initial infection), whereas if $R_0 > 1$, then the DFE is unstable.

Consider, for example, the SIR model (4.0.1). The DFE is $(A/\mu, 0, 0)$. In healthy population, one infection evolves according to

$$\frac{dI}{dt} = -\nu I - \mu I, \quad I(0) = 1,$$

so that $I(t) = e^{-(\nu+\mu)t}$ at time t, with total infection

$$\int_0^\infty I(t)dt = \frac{1}{\nu + \mu}.$$

The secondary infection is then

$$R_0 = \beta \frac{A}{\mu} \cdot \frac{1}{\nu + \mu}.$$

As already computed in (4.0.2) and (4.0.3), the DFE is stable if $R_0 < 1$ and unstable if $R_0 > 1$.

PROBLEM 4.1. Prove that if (4.0.2) holds, then DFE is globally stable, that is, for any initial condition, $(S, I, R) \to (A/\mu, 0, 0)$ as $t \to \infty$. [Hint: Note that $S(t) \le N(t)$, $N(t) \to A/\mu$ if $t \to \infty$, so that $S(t) < A/\mu + \epsilon'$ for arbitrarily small ϵ' if t is large enough, and then show that the function $x = I + \epsilon R$ satisfies $dx/dt \le -\delta x$ for some $\delta > 0$ and large t, if ϵ is positive and sufficiently small.]

PROBLEM 4.2. Prove that if (4.0.3) holds, then the equilibrium point in (4.0.4) is asymptotically stable. [Hint: Use the Routh–Hurwitz theorem.]

A stable equilibrium point with $I > 0$ is called **endemic**; it represents a disease that will never disappear. Thus, if (4.0.3) holds, then (4.0.4) represents an endemic disease.

When a susceptible is exposed to an infected individual he/she may or may not become immediately infected. With this in mind, we refine the SIR model by introducing a new class, E, of exposed individuals. The new model, called the **SEIR model**, consists of the following equations:

$$
\begin{aligned}
\frac{dS}{dt} &= A - \beta SI + \gamma R - \mu S, \\
\frac{dE}{dt} &= \beta SI - kE - \mu E, \\
\frac{dI}{dt} &= kE - \nu I - \mu I, \\
\frac{dR}{dt} &= \nu I - \gamma R - \mu R.
\end{aligned}
\tag{4.0.5}
$$

Here k is the rate by which the exposed become infected, and β is the rate by which a susceptible comes in contact with an infected individual.

The DFE for the SEIR model is $(A/\mu, 0, 0, 0)$.

PROBLEM 4.3. Show that the DFE of (4.0.5) is globally asymptotically stable if

$$\beta \frac{A}{\mu} < \frac{(\nu + \mu)(k + \mu)}{k}, \qquad (4.0.6)$$

that is, for any initial condition with $S(0) < \frac{A}{\mu}$, $(S, E, I, R) \to (\frac{A}{\mu}, 0, 0, 0)$ as $t \uparrow \infty$. [Hint: Show that the function

$$x \equiv E + \left(\frac{k + \mu}{k} - \epsilon \right) I + \epsilon R \quad (\epsilon > 0)$$

satisfies $dx/dt \le -\delta x$ for $\delta > 0$ and large t, if ϵ is positive and sufficiently small.]

PROBLEM 4.4. Prove that if the DFE of (4.0.5) is not stable, more precisely, if the inequality in (4.0.6) is reversed, then there exists another equilibrium point with $S \le \frac{A}{\mu}$.

In the SIR model we have taken the infection term to be βSI. Another possibility is to take the infection term to be $\beta SI/N$, where I/N is the relative proportion of the infected individuals, namely, the **prevalence** of the infection.

PROBLEM 4.5. (i) What is the DFE in (4.0.1) when βSI is replaced by $\beta SI/N$? Is it stable? (ii) If in the model with $\beta SI/N$ the DFE is not stable, does there exist an endemic equilibrium?

PROBLEM 4.6. Consider the SIR model with A replaced by the logistic growth $rN(1 - \frac{N}{K})$ where $N = S + I + R, r > \mu$. The DFE is $(K(1 - \frac{\mu}{r}), 0, 0)$:

(i) Under what conditions on the parameters is the DFE stable?
(ii) When the DFE is not stable, compute the nonzero steady state. Is it stable?

4.1. HIV Infection

In humans infected with HIV, the HIV viruses enter the CD4$^+$ T cells and hijack the machinery of the cells in order to multiply within these cells. As an infected T cell dies, an increased number of viruses emerge to invade and infect new CD4$^+$ T cells. This process eventually lead to significant depletion of the CD4$^+$ T cells, from over 700 in cm^3 of blood to 200 in cm^3. This state of the disease is characterized as AIDS; the immune system is too weak to sustain life for too long. In order to determine whether an initial infection with HIV will develop into AIDS we introduce a simple model which includes the CD4$^+$ T cells, denoted by T, the infected CD4$^+$ T cells, denoted by T^*, and the HIV viruses, denoted by V. Their number densities satisfy the following system of equations:

$$\begin{aligned}
\frac{dT}{dt} &= A - \beta TV - \mu T, \\
\frac{dT^*}{dt} &= \beta TV - \mu^* T^*, \\
\frac{dV}{dt} &= \gamma T^* - \nu V.
\end{aligned} \qquad (4.1.1)$$

Here β is the infection rate of a healthy T cell by external virus, μ and μ^* are the death rates of T and T^*, respectively, and γ is the number of virus particles that emerge upon death of one infected CD4$^+$ T cell.

PROBLEM 4.7. Prove that the DFE $(A/\mu, 0, 0)$ of (4.1.1) is asymptotically stable if and only if $\beta A/\mu < \nu\mu^*/\gamma$.

In model (4.1.1) we can compute the basic reproduction number R_0 as follows: One virus has the lifetime of $\frac{1}{\nu}$ (since $\frac{dV}{dt} = -\nu V$, $V(t) = e^{-\nu t}$, $\int_0^\infty V(t)dt = \frac{1}{\nu}$) and it infects A/μ T cells at rate β, and each infected T^* with lifetime $1/\mu^*$ gives rise to γ virus particles. Hence

$$R_0 = \frac{1}{\nu}\beta\frac{A}{\mu}\frac{1}{\mu^*}\gamma = \frac{\beta A \gamma}{\nu\mu\mu^*}.$$

From Problem 4.7 we see that the DFE is asymptotically stable if and only if $R_0 < 1$.

PROBLEM 4.8. Prove that $R_0 < 1$ if and only if all the eigenvalues of the Jacobian matrix of (4.1.1) about $(A/\mu, 0, 0)$ have negative real part.

4.2. Waterborne Disease

We next introduce a specific waterborne disease, namely, cholera. We introduce it as a simple SIR model with added compartment W that tracks pathogen concentration in water. The model is based on the following diagram:

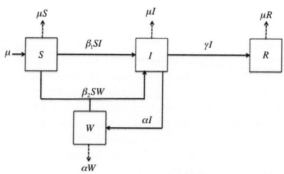

We assume that infected individuals do not die from the disease, so that the total population $N(t)$ is constant, and take $S + I + R = 1$. Based on the above diagram we then have the following equations:

$$
\begin{aligned}
\frac{dS}{dt} &= \mu - \beta_1 SI - \beta_2 SW - \mu S, \\
\frac{dI}{dt} &= \beta_1 SI + \beta_2 SW - \gamma I - \mu I, \\
\frac{dW}{dt} &= \alpha\,(I - W), \\
\frac{dR}{dt} &= \gamma I - \mu R.
\end{aligned}
\qquad (4.2.1)
$$

where β_2 is the water reservoir–person contact rate, β_1 is the person–person contact rate, $1/\gamma$ is the infectious period, $1/\alpha$ is the pathogen lifetime in the water reservoir, and μ is the birth/death rate. In the equation for W it was assumed that an infected individual contributes I pathogens through his/her waste product, and in water the lifetime of these pathogens is the same as the lifetime of the pathogens already in water. The phase space for the system is

$$
\Omega = \{(S, I, W, R)\,;\, S \geq 0, I \geq 0, 0 \leq W \leq I, R \geq 0, S + I + R = 1\}.
$$

Indeed, one can show that if $(S(0), I(0), W(0), R(0))$ belongs to Ω, then $(S(t), I(t), W(t), R(t))$ also belongs to Ω, for all $t > 0$. The DFE for the system is

$$
(S, I, W, R) = (1, 0, 0, 0) \qquad (4.2.2)
$$

and all the eigenvalues of the Jacobian matrix at $(1, 0, 0, 0)$ have negative real parts if and only if the eigenvalues of the matrix

$$
A = \begin{pmatrix} \beta_1 - \gamma - \mu & \beta_2 \\ \alpha & -\alpha \end{pmatrix}
$$

have negative real parts, i.e., if the trace is negative and the determinant of A is positive, that is, if

$$
\frac{\beta_1 + \beta_2}{\gamma + \mu} < 1.
$$

There is a procedure to compute the basic reproduction number for general disease models, which is beyond the scope of this book; see [20, Chap. 6] for general disease models. In the case of the model (4.2.1), one can then compute that $R_0 = (\beta_1 + \beta_2)/(\gamma + \mu)$.

The above mathematical model tells us that in order to control a cholera outbreak we need to make

$$
\beta_1 + \beta_2 < \gamma + \mu.
$$

Since the death rate μ is independent of the disease, interventions should aim either at increasing γ (by effective medical treatment), or at decreasing β_1 (contact between susceptible and infected individuals), or at decreasing β_2 (decontamination of water). There is a cost associated with each of these interventions, while some of these interventions could be taken jointly. Which is the optimal course of action should be determined by public health authorities, taking into consideration local conditions.

4.3. Numerical Methods for Ordinary Differential Equations

4.3.1. Euler Method. In Sect. 2.1.3, we discussed how to solve ordinary differential equations

$$\frac{dx_i}{dt} = f_i(x_1, \ldots, x_n) \quad (1 \leq i \leq n)$$

by using the built-in MATLAB function "dsolve." However, the ODEs describing a biological system may not have closed-form solutions. In this case, one can look for numerical approaches [15, 4] that provide approximation of the exact solutions. Since numerical solutions are not exact, it is important to understand how accurate numerical solutions are and how robust the numerical approaches are.

We start with the simplest method: Euler method, which is named after Leonhard Euler (1707–1783); notice that this method was introduced much earlier before computers were invented. Nowadays, it is one of the methods that people who work on computational mathematics must learn.

Suppose the system of ODEs we want to solve is

$$\dot{\mathbf{x}} = \mathbf{f}(\mathbf{x}, t), \quad t \geq t_0, \quad \mathbf{x}(t_0) = \mathbf{x}_0 \tag{4.3.1}$$

where $\mathbf{f}$ is Lipschitz function in $\mathbf{x}$ which maps $R^d \times [t_0, \infty)$ to R^d and the initial condition $\mathbf{x}_0$ is a given point in R^d. There are many different ways to derive Euler method. Here we give one derivation based on a linear interpolant. Starting with $\mathbf{x}(t_0) = \mathbf{x}_0$, $\mathbf{x}(t)$ is estimated by making the approximation $\mathbf{f}(\mathbf{x}(t), t) \approx \mathbf{f}(\mathbf{x}(t_0), t_0)$ for t near t_0. Thus

$$\mathbf{x}(t) = \mathbf{x}(t_0) + \int_{t_0}^{t} \mathbf{f}(\mathbf{x}, \tau) d\tau \approx \mathbf{x}_0 + (t - t_0)\mathbf{f}(\mathbf{x}(t_0), t_0).$$

If t is sufficiently close to t_0, this should provide a good approximation. Introducing h as the step size, $h > 0$, we then define the numerical solution by

$$\mathbf{X}(t_0 + h) = \mathbf{X}_0 + h\mathbf{f}(\mathbf{X}(t_0), t_0), \quad \text{where} \quad \mathbf{X}_0 = \mathbf{x}_0.$$

Repeating this process for a sequence $t_n = t_0 + nh$ yields the celebrated Euler method

$$\mathbf{X}_{n+1} = \mathbf{X}_n + h\mathbf{f}(\mathbf{X}(t_n), t_n) \tag{4.3.2}$$

where $\mathbf{X}_n = \mathbf{X}(t_0 + nh)$ is a numerical approximation solution of $\mathbf{x}(t_0 + nh)$. The numerical scheme (4.3.2) can be written in the form

$$\mathbf{X}_{n+1} = \mathbf{H}(\mathbf{X}_n)$$

where $\mathbf{H}(\mathbf{X}_n) = \mathbf{X}_n + h\mathbf{f}(\mathbf{X}(t_n), t_n)$. This type of scheme is called **explicit scheme** because the solution $\mathbf{X}_{n+1}$ is explicitly defined as a function of $\mathbf{X}_n$. Furthermore, it is a **single step** method because it requires the solution at only one previous time step to obtain the solution at the current step. We will revisit this concept when we discuss more complicated schemes later.

In order to understand how good the numerical solution is, one can introduce the **local truncation error** by

$$L_{TE} = \mathbf{x}_{n+1} - \mathbf{H}(\mathbf{x}_n).$$

For Euler method, if $\mathbf{f}(\mathbf{x}, t)$ has two continuous derivatives in $\mathbf{x}$, then

$$L_{TE} = \mathbf{x}_{n+1} - \mathbf{H}(\mathbf{x}_n) = \mathbf{x}_{n+1} - \mathbf{x}_n - h\mathbf{f}(\mathbf{x}(t_n), t_n) = \frac{h^2}{2}\mathbf{x}_n''(\bar{t}_n) = O(h^2)$$

where $\bar{t}_n$ is some point in the interval $[t_n, t_n + h]$. If a method has the local truncation

$$L_{TE} = O(h^{p+1}) \tag{4.3.3}$$

where $p \geq 1$, we say that the method is **consistent**. Thus Euler method is consistent if $\mathbf{f}(\mathbf{x}, t)$ has two continuous derivatives in $\mathbf{x}$.

DEFINITION. Euler method is said to be **convergent** if $\max_n |e_{n,h}| \to 0$ as $h \to 0$, where

$$e_{n,h} = \mathbf{x}(t_0 + nh) - \mathbf{X}(t_0 + nh).$$

Next we demonstrate that Euler method is indeed convergent.

THEOREM 4.1. *If $\mathbf{f}(x)$ has two continuous derivatives, Euler method is convergent.*

PROOF. Using Taylor's expansion, the exact solution satisfies

$$\mathbf{x}(t_{n+1}) = \mathbf{x}(t_n) + h\mathbf{x}'(t_n) + O(h^2) = \mathbf{x}(t_n) + h\mathbf{f}(\mathbf{x}(t_n), t_n) + O(h^2) \tag{4.3.4}$$

where $|O(h^2)| \leq c_2 h^2$. Subtracting (4.3.2) from (4.3.4) gives

$$e_{n+1,h} = e_{n,h} + h\left[\mathbf{f}(\mathbf{X}(t_n) + e_{n,h}, t) - \mathbf{f}(\mathbf{X}(t_n), t)\right] + O(h^2)$$

Since $\mathbf{f}$ is Lipschitz continuous, we then get

$$|e_{n+1,h}| \leq |e_{n,h}| + hc_1|e_{n,h}| + c_2 h^2 \tag{4.3.5}$$

where c_1 is the Lipschitz constant. Using induction one can deduce from (4.3.5) the inequalities

$$|e_{n,h}| \leq \frac{c_2}{c_1}h[(1 + hc_1)^n - 1], \quad n = 0, 1, 2, \dots \quad . \tag{4.3.6}$$

Suppose $nh = T$ is the final time we are interested in. Since

$$(1 + hc_1)^n \leq e^{nhc_1} \leq e^{Tc_1},$$

it follows from (4.3.6) that

$$|e_{n,h}| \leq \frac{c_2}{c_1}h[e^{Tc_1} - 1]. \tag{4.3.7}$$

Noting that $\frac{c_2}{c_1}[e^{Tc_1} - 1]$ is independent of h, we conclude that

$$\lim_{h \to 0} |e_{n,h}| = 0.$$

$\square$

The error bound in (4.3.7) contains an exponential term which, usually, does not provide a sharp error bound, especially when c_1 or T is large. For example, consider the equation

$$x' = -100x \quad \text{with} \quad x(0) = 1; \tag{4.3.8}$$

the exact solution is $x(t) = e^{-100t}$. We shall now apply Euler method to this equation. The Lipschitz constant is $c_1 = 100$ which gives a "very bad" bound in (4.3.7). Euler method produces the numerical approximation

$$X(nh) = (1 - 100h)^n.$$

The numerical error is

$$|e_{n,h}| = |e^{-100nh} - (1 - 100h)^n|$$

which is smaller by many orders of magnitude than the bound on the right-hand side of (4.3.7). Even though the bound (4.3.7),

$$|e_{n,h}| = O(h),$$

seems useless, it does provide a useful information, namely, that the error decreases by half if the step size is halved. When a numerical method has this latter property, it is said to be of **first-order accuracy**. In general, if

$$|e_{n,h}| = O(h^p),$$

we say the method is p-**order accurate**. When the exact solution is not available, we can extract numerical order by comparing the numerical results at the same final time T from different step sizes

$$
\begin{aligned}
\frac{|X_{2h}(T) - X_{4h}(T)|}{|X_h(T) - X_{2h}(T)|} &= \frac{|(X_{2h}(T) - x(T)) - (X_{4h}(T) - x(T))|}{(X_h(T) - x(T)) - (X_{2h} - x(T))|} \\
&= \frac{|e_{2h} - e_{4h}|}{|e_h - e_{2h}|} \approx \frac{|c(2h)^p - c(4h)^p|}{c(h)^p - c(2h)^p} = 2^p
\end{aligned}
$$

Thus the order p can be extracted by the formula

$$p = \log_2\left(\frac{|X_{2h}(T) - X_{4h}(T)|}{|X_h(T) - X_{2h}(T)|}\right).$$

Another interesting phenomenon to observe in the special example of (4.3.8) is that Euler method generates a monotone decreasing sequence X_n and the convergent occurs when $|1 - 100h| < 1$. It follows that the step size needs to be small, namely,

$$h < \frac{1}{50},$$

or else the approximate solution will oscillate and diverge.

In order to come up with schemes which are more stable, one resorts to implicit-type schemes. The implicit Euler scheme is the simplest scheme in this family. The update formula is

$$\mathbf{X}_{n+1} = \mathbf{X}_n + h\mathbf{f}(\mathbf{X}(t_{n+1}), t_{n+1}) \tag{4.3.9}$$

and it can be written in the form of

$$\mathbf{X}_{n+1} = \mathbf{H}(\mathbf{X}_n, \mathbf{X}_{n+1})$$

where $\mathbf{H}(\mathbf{X}_n) = \mathbf{X}_n + h\mathbf{f}(\mathbf{X}(t_{n+1}), t_{n+1})$. This type of scheme is called **implicit scheme** because the solution $\mathbf{X}_{n+1}$ is implicitly defined as a function of $\mathbf{X}_n$.

PROBLEM 4.9. Deduce (4.3.6) from (4.3.5).

PROBLEM 4.10. Write a code to implement both explicit and implicit Euler methods for the system of Eqs. (2.0.7)–(2.0.9) with $k_1 = 0.1$, $k_{-1} = 1.e - 5$, $k_2 = 1$. Choose the initial condition $[P] = [E] = [S] = 1$ and vary $[C] = 1 : 1 : 6$. Plot the solution and describe what you have observed in the numerical results. Indicate your step size.

4.3.2. Runge–Kutta Methods.
Even though explicit (forward) and implicit (backward) Euler methods are easy to implement, they are both first-order methods. In order to achieve highly accurate results, the step size needs to be small. In this section, we introduce a family of methods called Runge–Kutta (RK) methods which can generate more accurate results.

Runge–Kutta methods were developed by the mathematicians C. Runge and M.W. Kutta. Similar to Euler methods, both explicit and implicit RK methods were introduced. The main idea of RK method is based on a quadrature rule. Integrating the Eq. (4.3.1) from $t_n = nh$ to $t_{n+1} = (n+1)h$ gives

$$\mathbf{x}(t_{n+1}) = \mathbf{x}(t_n) + \int_{t_n}^{t_{n+1}} \mathbf{f}(\mathbf{x}, \tau)d\tau = \mathbf{x}(t_n) + h\int_0^1 \mathbf{f}(\mathbf{x}(t_n + h\tau), t_n + h\tau)d\tau.$$

The integral term is then approximated by the quadrature rule

$$\int_0^1 \mathbf{f}(\mathbf{x}(t_n + h\tau), t_n + h\tau)d\tau \approx \sum_{j=1}^{\nu} b_j f(\mathbf{x}(t_n + c_j h), t_n + c_j h) \qquad (4.3.10)$$

where $\mathbf{b} = (b_j)_{j=1,\ldots,\nu}$ are RK weights and $\mathbf{c} = (c_j)_{j=1,\ldots,\nu}$, are RK nodes with $0 \le c_1 \le c_2 \le \ldots \le c_\nu \le 1$. Note that it is not necessary to have distinct nodes. When $\nu = 1$, we choose $c_1 = 0$ and $b_1 = 1$ ($c_1 = 1$ and $b_1 = 1$), and the method goes back to explicit (implicit) Euler method. If $\nu = 2$, $c_1 = 0$, $c_2 = 1$, and $b_1 = b_2 = \frac{1}{2}$, the method is called trapezoidal method. If the c_j are neither 0 nor 1, in order to build an implementable scheme, the intermediate values $\mathbf{x}(t_n + c_j h)$ need to be further approximated. The explicit RK method uses the approximation

$$\xi_j = \mathbf{x}_n + h\sum_{i=1}^{j-1} a_{j,i} f(\xi_i, t_n + c_i h), \quad j = 1, 2, \ldots, \nu, \qquad (4.3.11)$$

where $\xi_j = \mathbf{x}(t_n + c_j h)$, while the implicit RK method uses the approximation

$$\xi_j = \mathbf{x}_n + h \sum_{i=1}^{\nu} a_{j,i} f(\xi_i, t_n + c_i h), \quad j = 1, 2, \ldots, \nu; \tag{4.3.12}$$

in both methods the next time step is defined by

$$\mathbf{x}_{n+1} = \mathbf{x}_n + h \sum_{j=1}^{\nu} b_j f(\xi_j, t_n + c_j h), \tag{4.3.13}$$

The matrix $A = (a_{j,i})_{j,i=1,2,\ldots,\nu}$, where missing elements are defined to be zero, is called the RK matrix. Notice that the A matrix of explicit RK methods is strictly lower triangular.

The determination of RK weights $\mathbf{b}$, nodes $\mathbf{c}$, and matrix A depends on the order of accuracy required. The explicit Runge–Kutta method is consistent [recall (4.3.3)] if

$$\sum_{i=1}^{j-1} a_{j,i} = c_j \quad \text{for} \quad j = 2, 3, \ldots, \nu.$$

Here we give a simple example with $\nu = 2$ and $c_1 = 0$ for the second-order explicit RK method, i.e.,

$$\begin{aligned}
\mathbf{x}_{n+1} &= \mathbf{x}_n + h\left(b_1 \mathbf{f}\left(\xi_1, t_n\right) + b_2 \mathbf{f}\left(\xi_2, t_n + c_2 h\right)\right), \\
\xi_1 &= \mathbf{x}_n, \\
\xi_2 &= \mathbf{x}_n + h a_{2,1} \mathbf{f}(\xi_1, t_n),
\end{aligned}$$

The local truncation error is

$$\begin{aligned}
L_{TE} = \mathbf{x}(t_n + h) - \mathbf{x}(t_n) - h\,(b_1 \mathbf{f}\left(\mathbf{x}(t_n), t_n\right) + b_2 \mathbf{f}\left(\mathbf{x}(t_n)\right. \\
\left. + h a_{2,1} \mathbf{f}\left(\mathbf{x}(t_n), t_n\right), t_n + c_2 h\right)).
\end{aligned}$$

Using Taylor's expansion yields

$$\begin{aligned}
L_{TE} &= h\mathbf{x}'(t_n) + \tfrac{h^2}{2}\mathbf{x}''(t_n) - h\left((b_1 + b_2)\,\mathbf{f}(\mathbf{x}(t_n), t_n)\right) \\
&\quad - h^2\left(b_2 c_2 \tfrac{\partial f(x(t_n), t_n)}{\partial t} + a_{2,1} b_2 \tfrac{\partial f(x_n, t_n)}{\partial x} f(x_n, t_n)\right) + O(h^3).
\end{aligned}$$

Substituting the equation $x'(t_n) = f(x(t_n), t_n)$ and $x''(t_n) = f_t(x(t_n), t_n) + f_x(x(t_n), t_n) f(x_n, t_n)$, the local truncation error becomes

$$\begin{aligned}
L_{TE} &= \tfrac{h^2}{2}\mathbf{x}''(t_n)h\left((1 - b_1 - b_2)\,\mathbf{f}(\mathbf{x}(t_n), t_n)\right) \\
&\quad - h^2\left(\left(\tfrac{1}{2} - b_2 c_2\right)\tfrac{\partial f(x(t_n), t_n)}{\partial t} + \left(\tfrac{1}{2} - a_{2,1} b_2\right)\tfrac{\partial f(x_n, t_n)}{\partial x} f(x_n, t_n)\right) \\
&\quad + O(h^3).
\end{aligned}$$

If we require the method to be of second-order accuracy, RK weights $\mathbf{b}$, nodes $\mathbf{c}$, and matrix A must satisfy

$$
\begin{aligned}
b_1 + b_2 &= 1 \\
b_2 c_2 &= \tfrac{1}{2} \\
a_{2,1} b_2 &= \tfrac{1}{2}
\end{aligned}
\qquad (4.3.14)
$$

The conditions (4.3.14) do not define a unique set of parameters. Three popular choices of parameters displayed in the **RK tableaux form**

$$
\begin{pmatrix}
c_1 & a_{1,1} & \cdots & a_{1,\nu} \\
\vdots & \vdots & \ddots & \vdots \\
c_\nu & a_{\nu,1} & \cdots & a_{\nu,\nu} \\
 & b_1 & \cdots & b_\nu
\end{pmatrix}
$$

are

$$
\begin{pmatrix}
0 & & \\
\tfrac{1}{2} & \tfrac{1}{2} & \\
 & 0 & 1
\end{pmatrix},
\begin{pmatrix}
0 & & \\
\tfrac{2}{3} & \tfrac{2}{3} & \\
 & \tfrac{1}{4} & \tfrac{3}{4}
\end{pmatrix},
\text{ and }
\begin{pmatrix}
0 & & \\
1 & 1 & \\
 & \tfrac{1}{2} & \tfrac{1}{2}
\end{pmatrix}.
$$

The higher-order explicit RK method can be derived in the similar way but the calculation is more involved. The conditions for third-order accuracy are

$$
\begin{aligned}
b_1 + b_2 + b_3 &= 1 \\
b_2 c_2 + b_3 c_3 &= 1 \\
b_2 c_2^2 + b_3 c_3^2 &= \tfrac{1}{3} \\
b_3 a_{3,2} c_2 &= \tfrac{1}{6}
\end{aligned}
\qquad (4.3.15)
$$

Two most common third-order explicit RK schemes are

$$
\begin{pmatrix}
0 & & & \\
\tfrac{1}{2} & \tfrac{1}{2} & & \\
1 & -1 & 2 & \\
 & \tfrac{1}{6} & \tfrac{2}{3} & \tfrac{1}{6}
\end{pmatrix},
\text{ and }
\begin{pmatrix}
0 & & & \\
\tfrac{2}{3} & \tfrac{2}{3} & & \\
\tfrac{2}{3} & 0 & \tfrac{2}{3} & \\
 & \tfrac{1}{4} & \tfrac{3}{8} & \tfrac{3}{8}
\end{pmatrix}.
$$

For fourth-order explicit RK scheme, the most well-known tableaux form is

$$
\begin{pmatrix}
0 & & & & \\
\tfrac{1}{2} & \tfrac{1}{2} & & & \\
\tfrac{1}{2} & 0 & \tfrac{1}{2} & & \\
1 & 0 & 0 & 1 & \\
 & \tfrac{1}{6} & \tfrac{1}{3} & \tfrac{1}{3} & \tfrac{1}{6}
\end{pmatrix}
$$

The implicit family of RK methods is more difficult to implement. However, it is much more stable and can be implemented with larger time steps. We skip the discussion on implicit RK methods because they are beyond the scope of this book. The interested reader is referred to [15].

4.3.3. MATLAB Built-In ODE Solvers. The built-in function in MATLAB based on Runge–Kutta 4th- and 5th-order methods [8] to solve the system of ODEs (4.3.1) is

[T,Y] = ode45(odefun,tspan,y0)

with an input tspan = [t0 tf] which integrates the system of differential equations specified by odefun from time t0 to tf with initial conditions y0. Let us try to solve the ODE system

$$
\begin{array}{rclr}
x_1' & = & x_2 x_3 & x_1(0) = 0, \\
x_2' & = & -x_1 x_3 & x_2(0) = 1, \\
x_3' & = & -0.51 x_1 x_2 & x_3(0) = 1,
\end{array}
$$

which was proposed by Krogh. To solve this system, first create a function ODEsym1 containing the differential equations

```
function dx = ODEsym1(t,x)
dx = zeros(3,1); % a column vector
dx(1) = x(2) * x(3);
dx(2) = -x(1) * x(3);
dx(3) = -0.51 * x(1) * x(2);
```

In this demo example shown in MATLAB instruction manual, the error tolerances were changed using the odeset command and the system is solved on a time interval [0 12] with an initial condition vector [0 1 1] at time 0:

```
>>options = odeset('RelTol',1e-4,'AbsTol',[1e-4 1e-4 1e-5]);
>>[T,X] = ode45(@ODEsym1,[0 12],[0 1 1],options);
```

Plotting the columns of the returned array X versus T shows the solution in Fig. 4.1:

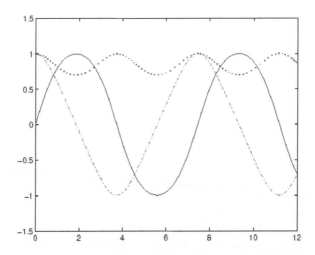

FIGURE 4.1. The plot of the solutions of ODEsym1

>>plot(T,X(:,1),'-',T,X(:,2),'-.',T,X(:,3),'.')

PROBLEM 4.11. Derive the necessary conditions for the third-order explicit RK scheme.

PROBLEM 4.12. Prove that the order of two-stage (i.e., $\nu = 2$) explicit RK method cannot exceed two.

PROBLEM 4.13. Choose MATLAB built-in solver or an explicit RK method discussed in this chapter to solve the SEIR model (4.0.5) with different β to show DFE of (4.0.5) can be either stable or unstable. Demonstrate the order of accuracy numerically and list the parameters you choose.

CHAPTER 5

Chemostats and Competition Among Species

The chemostat, or bioreactor, is a continuous stirred-tank reactor (CSTR) used for continuous production of microbial biomass. It consists of a freshwater and nutrient reservoir connected to a growth chamber, or reactor, with microorganism. The mixture of freshwater and nutrient is pumped continuously from the reservoir to the reactor chamber, providing feed to the microorganism, and the mixture of culture and fluid in the growth chamber is continuously pumped out and collected. The medium culture is continuously stirred. Stirring ensures that the contents of the chamber are well mixed so that the culture production is uniform and steady. If the steering speed is too high, it would damage the cells in culture, but if it is too low, it could prevent the reactor from reaching steady-state operation. Figure 5.1 is a conceptual diagram of a chemostat.

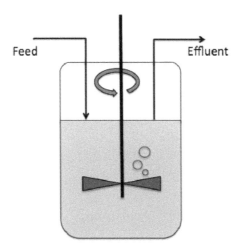

FIGURE 5.1. Stirred bioreactor operated as a chemostat, with a continuous inflow (the feed) and outflow (the effluent). The rate of medium flow is controlled to keep the culture volume constant

© Springer International Publishing Switzerland 2014

A. Friedman, C.-Y. Kao, *Mathematical Modeling of Biological Processes*, Lecture Notes on Mathematical Modelling in the Life Sciences, DOI 10.1007/978-3-319-08314-8_5

Chemostats are used to grow, harvest, and maintain desired cells in a controlled manner. The cells grow and replicate in the presence of suitable environment with medium supplying the essential nutrient growth. Cells grown in this manner are collected and used for many different applications.

The **dilution rate**, D, is defined as the rate of flow of the medium in the growth chamber over the volume of the culture in the bioreactors:

$$D = \frac{\text{medium flow rate}}{\text{culture volume}} = \frac{F}{V}.$$

If the dilution rate is chosen too large, the culture will not be able to sustain itself in the bioreactor since the cells have a limited growth rate; this will result in a washout. On the other hand, if the dilution rate is too small, the production of the culture is then also too slow, and the chemostat is not working efficiently. So the question for the chemostat designer is how to determine the optimal rate so that the chemostat works, efficiently and safely, in steady state. This question can be addressed by mathematical modeling of chemostat.

Figure 5.2 shows a chemostat with an overflow system: the level of the culture is maintained by overflow of the effluent through a port on the side of the reactor. The rates of the feed and effluent are the same.

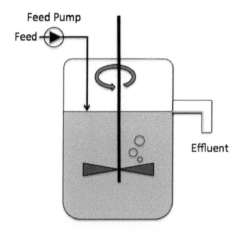

FIGURE 5.2. Chemostat with an overflow system

A bacterial chemostat is a specific type of bioreactor, and its design may depend on the specific application of the bacterial product. These applications include:

Pharmaceutical: for example in analyzing how bacteria respond to different antibiotics or in production of insulin (by the bacteria) for diabetics

Food industry: for production of fermented food such as cheese

Manufacturing: for fermenting sugar to produce ethanol

Chemostats are also used in research for investigations in cell biology, as a source for large volumes of uniform eukaryotic cells or bacteria. The chemostat is often used to gather steady-state data about an organism in order to develop models regarding the metabolic processes within the organism. Such models aim to study what will be the effect of induced mutation on a bacteria, how a pathogen will react to antibiotic drug, or how a bacteria will utilize a mixture of two different sources of nutrients. Competition among several populations of bacteria, some of which may be predators, is another important area of research. An important question, for instance, is which of several bacterial strains will survive and which will become extinct in a competition for resources. Such a question may be addressed by mathematical models, as will be shown in this chapter.

Figure 5.3 is a schematics of a chemostat used for our mathematical model. A stock of nutrients with concentration C_0 is pumped into the chamber of the bacterial culture, and an outflow valve allows the culture to be pumped out. We introduce the following notation:

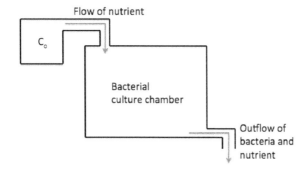

FIGURE 5.3. Schematics of a chemostat

$$
\begin{aligned}
V &= \text{volume of the bacterial culture chamber,} \\
C(t) &= \text{concentration of nutrients in the chamber,} \\
F &= \text{rate of inflow and outflow,} \\
x &= \text{concentration of the bacteria in the chamber.}
\end{aligned}
$$

We assume that
$$
\frac{\text{mass of the bacteria formed}}{\text{mass of the nutrients used}} = const. = \gamma;
$$
γ is called the **yield constant**. By conservation of nutrient mass,

$$\text{rate of change} = \text{input-washout-consumption.}$$

Based on experimental evidence we take the bacterial load to be as in the Michaelis–Menten law
$$
\frac{m_0 C}{a + C} x
$$

where m_0 and a are constants and the rate of nutrient consumption to be

$$\frac{m_0 C}{a + C} \frac{x}{\gamma},$$

since mass $1/\gamma$ of the bacteria is formed from consumption of mass 1 of nutrients. Then

$$(VC)'(t) = C_0 F - C(t)F - \frac{m_0 C}{a + C} \frac{x}{\gamma}.$$

Dividing both sides by V and setting $D = F/V$ (the **dilution rate**), we get

$$C' = (C_0 - C)D - \frac{mC}{a + C} \frac{x}{\gamma} \tag{5.0.1}$$

where $m = m_0/V$. The bacterial growth is given by

$$x' = x \left(\frac{mC}{a + C} - D \right). \tag{5.0.2}$$

Note that the units of C_0, C, a, x are mass/volume (e.g., gm/cm^3) and the units of m and D are $1/time$ (e.g., $1/s$); γ is a dimensionless parameter.

By scaling

$$\bar{C} = \frac{C}{C_0}, \quad \bar{x} = \frac{x}{\gamma C_0}, \quad \bar{t} = Dt$$

we can simplify the system (5.0.1) and (5.0.2). After dropping the bars over C and x we then obtain (with new constants $\bar{m} = \frac{m}{D}$, $\bar{a} = \frac{a}{C_0}$):

$$
\begin{aligned}
C' &= 1 - C - \frac{\bar{m}Cx}{\bar{a}+C}, \\
x' &= x \left(\frac{\bar{m}C}{\bar{a}+C} - 1 \right).
\end{aligned}
\tag{5.0.3}
$$

PROBLEM 5.1. The steady states of (5.0.3) are $(C_1, x_1) = (1, 0)$ and $(C_2, x_2) = (\lambda, 1 - \lambda)$ where $\lambda = \frac{\bar{a}}{\bar{m}-1}$, provided $\bar{m} > 1, \lambda < 1$. Note that $\frac{\bar{m}}{\bar{a}+1} > 1$ if and only if $\lambda < 1$. Prove

(i) (C_1, x_1) is asymptotically stable if $\frac{\bar{m}}{\bar{a}+1} < 1$.

(ii) (C_2, x_2) is asymptotically stable.

To biologically interpret the mathematical results of Problem 5.1 we return to the original parameters and consider for example the role of the dilution rate D. Setting

$$D_0 = \frac{m_0/V}{a/C_0 + 1},$$

we have

$$\frac{\bar{m}}{\bar{a}+1} = \frac{(m_0/V)/D}{a/C_0 + 1} = \frac{D_0}{D}.$$

If $D > D_0$, then $\bar{m}/(\bar{a}+1) < 1$, so that $(C_1, x_1) = (1, 0)$ is stable, and in steady state the chemostat does not produce any bacteria, that is, if $D > D_0$, then there is a washout. On the other hand, if $D < D_0$, then $\bar{m}/(\bar{a}+1) > 1$, so that $\bar{m} > 1$ and $\lambda < 1$; hence, in steady state, the chemostat yields bacteria at the (scaled) amount $1 - \lambda$, and one can adjust the parameter D, or other parameters of the model, to obtain the desired amount of bacteria per nutrient.

Since $\bar{t} = Dt$, we have

$$\frac{dx}{dt} = \frac{dx}{d\bar{t}} \frac{d\bar{t}}{dt} = D\frac{dx}{d\bar{t}}$$

for the effluent x. Hence in steady state when the bacterial yield is $1 - \lambda$ in unit time $\bar{t}$, the actual bacterial yield per unit time (when $D < D_0$) is

$$D\left(1 - \lambda\right) = D\left(1 - \frac{\bar{a}}{\bar{m} - 1}\right) = D\left(1 - \frac{aVD}{C_0(m_0 - VD)}\right) \equiv f(D).$$

Hence, to maximize the bacterial harvest, one should take the dilution rate D such that $f(D)$ becomes maximum in the interval $0 < D < D_0$.

PROBLEM 5.2. Prove that the maximum of $f(D)$ is attained at the smaller of the two positive solutions of the quadratic equation

$$\alpha D^2 + \beta D + m_0^2 = 0,$$

where $\alpha = V(M + V)$, $\beta = -2m_0(M + V)$, $M = aV/C_0$.

PROBLEM 5.3. Prove the following statements and state, for parts (ii)–(iv), their biological conclusions in terms of the original model parameters:

(i) $C(t) + x(t) \to 1$ as $t \to \infty$.
(ii) If $\bar{m} < 1$, then $x(t) \to 0$ as $t \to \infty$.
  In parts (iii)–(iv) assume that $C(0) + x(0) = 1$ so that $C(t) + x(t) = 1$ for all $t > 0$.
(iii) If $\bar{m} > 1$ and $\lambda \equiv \frac{\bar{a}}{\bar{m}-1} > 1$, then $x(t) \to 0$ as $t \to \infty$.
(iv) If $\bar{m} > 1$ and $\lambda < 1$, then $\lim_{t\to\infty} x(t) = 1 - \lambda$, provided $x(0) > 0$.

Consider next a situation where two different microorganisms are in the chemostat, x_1 and x_2, and denote the corresponding parameters $\bar{a}$ and $\bar{m}$ by a_i, m_i. Then the chemostat equations become

$$\begin{aligned} C' &= 1 - C - \frac{m_1 C x_1}{a_1 + C} - \frac{m_2 C x_2}{a_2 + C}, \\ x_i' &= x_i\left(\frac{m_i C}{a_i + C} - 1\right) \quad (i = 1, 2). \end{aligned} \tag{5.0.4}$$

Set

$$\lambda_i = \frac{a_i}{m_i - 1},$$

and assume that $\lambda_1 \neq \lambda_2$. As in Problem 5.1(i),

$$C(t) + x_1(t) + x_2(t) \to 1 \quad \text{as} \quad t \to \infty.$$

Suppose $C(0) + x_1(0) + x_2(0) = 1$. Then

$$C(t) + x_1(t) + x_2(t) \equiv 1 \quad \text{for all } t > 0,$$

and (5.0.4) reduces to

$$x_i' = x_i \frac{m_i - 1}{1 + a_i - x_1 - x_2} (1 - \lambda_i - x_1 - x_2) \quad (i = 1, 2). \tag{5.0.5}$$

This system has three steady states:

$$E_0 = (0,0), \quad E_1 = (1 - \lambda_1, 0), \quad E_2 = (0, 1 - \lambda_2).$$

PROBLEM 5.4. Suppose that $m_1 > 1$, $m_2 > 1$, $0 < \lambda_1 < \lambda_2 < 1$. Find which of the steady states E_i is stable and provide biological interpretation.

The system (5.0.4) is an example of two populations competing for resources. Another example where the two populations are in conflict, killing each other, is the **Lotka–Volterra** system

$$
\begin{aligned}
\frac{dx}{dt} &= r_1 x \left(1 - \frac{x}{k_1}\right) - b_1 xy, \\
\frac{dy}{dt} &= r_2 y \left(1 - \frac{y}{k_2}\right) - b_2 xy.
\end{aligned}
\tag{5.0.6}
$$

It has four steady states:

$$E_1 = (0,0), \quad E_2 = (k_1, 0) \quad E_3 = (0, k_2)$$

$$E_4 = \left(\frac{\beta_1 k_2 - k_1}{\beta_1 \beta_2 - 1}, \frac{\beta_2 k_1 - k_2}{\beta_1 \beta_2 - 1}\right) \quad \text{where} \quad \beta_i = \frac{k_i b_i}{r_i}.$$

E_4 is biologically relevant only if

$$k_2 > \frac{r_1}{b_1}, \quad k_1 > \frac{r_2}{b_2}$$

or

$$k_2 < \frac{r_1}{b_1}, \quad k_1 < \frac{r_2}{b_2}.$$

PROBLEM 5.5. Determine which of the steady states E_i is stable, and give biological explanation.

Predator–prey models look somewhat like models of competing populations. The only difference is that one species feeds on another but not vice versa. Consider for example two competing predators, x and y, and one prey, z:

$$
\begin{aligned}
\frac{dx}{dt} &= r_1 x \left(1 - \frac{x}{k_1}\right) - b_1 xy + \beta_1 xz, \\
\frac{dy}{dt} &= r_2 y \left(1 - \frac{y}{k_2}\right) - b_2 xy + \beta_2 yz, \\
\frac{dz}{dt} &= r_3 z \left(1 - \frac{z}{k_3}\right) - \gamma_1 xz - \gamma_2 yz.
\end{aligned}
\tag{5.0.7}
$$

PROBLEM 5.6. Find the equilibrium points $(\bar{x}, \bar{y}, \bar{z})$ of the system (5.0.7) and determine their stability in the following cases: (i) $\bar{x} = 0, \bar{y} > 0, \bar{z} = 0$; (ii) $\bar{x} > 0, \bar{y} > 0, \bar{z} > 0$ and $b_1 = b_2 = 0$; you may need to use the Routh–Hurwitz theorem.

Consider the predator–prey model

$$
\begin{aligned}
\frac{du}{dt} &= ru(u - \alpha)(1 - u) - \frac{uv}{1 + \beta u} \qquad (u = \text{prey}), \\
\frac{dv}{dt} &= \frac{uv}{1 + \beta u} - \delta v \qquad\qquad (v = \text{predator}),
\end{aligned}
\tag{5.0.8}
$$

where $r > 0, \beta > 0, 0 < \alpha < 1, 0 < \delta < \frac{1}{1+\beta}$. The prey is said to be subject to the **Allee effect** in the sense that its growth rate $r(u - \alpha)(1 - u)$ is negative for small population u (i.e., if $u < \alpha$); endangered species are subject to the Allee effect.

PROBLEM 5.7. Consider the system (5.0.8). (i) Prove that the steady state $(0,0)$ is asymptotically stable, i.e., both populations will go extinct if each population is initially small. (ii) Prove that there is a unique steady state $(\bar{u}, \bar{v})$ with $\bar{u} > 0, \bar{v} > 0$ if $\delta > \frac{\alpha}{1+\beta\alpha}$, and that it is asymptotically stable if $\delta > \frac{1+\alpha}{2}$ and β is sufficiently small.

PROBLEM 5.8. Consider the model of one predator x and two prey species y and z:

$$\frac{dx}{dt} = \beta_1 xy + \beta_2 xz - \lambda x$$

$$\frac{dy}{dt} = r_1 y - \gamma_1 xy$$

$$\frac{dz}{dt} = r_2 z(1 - z) - \gamma_2 xz.$$

Check that the only steady point $(\bar{x}, \bar{y}, \bar{z})$ with $\bar{x} > 0, \bar{y} > 0, \bar{z} > 0$ is given by

$$\bar{x} = \frac{r_1}{\gamma_1}, \quad \bar{z} = 1 - \frac{\gamma_2}{r_2}\bar{x}, \quad \beta_1 \bar{y} = \lambda - \beta_2 \bar{z}$$

provided $\gamma_2 \bar{x} < r_2$ and $\beta_2 \bar{z} < \lambda$. Use the Routh–Hurwitz theorem to prove that $(\bar{x}, \bar{y}, \bar{z})$ is asymptotically stable.

5.1. Food Chain

The next example is a simple model of a food chain. We consider a predator–prey model in which the prey feeds on nutrients that are recycled through secretion and death of both prey and predator. The nutrient is nitrogen in the ocean, x, the prey is phytoplankton y, and predator is zooplankton z which consumes only the prey. Then

$$\frac{dx}{dt} = \alpha y + \beta z - \gamma xy,$$

$$\frac{dy}{dt} = \gamma xy - \delta yz - \alpha y,$$

$$\frac{dz}{dt} = \delta yz - \beta z,$$

where αy and βz are the recycled nitrogen through secretion and death of y and z, respectively, γy is the rate of nitrogen consumption by phytoplankton, and δz is the rate of phytoplankton consumption by zooplankton. The relation

$$\frac{dx}{dt} + \frac{dy}{dt} + \frac{dz}{dt} = 0 \quad \text{or} \quad x + y + z = const. = A$$

can be viewed as conservation of the total nitrogen mass.

The steady point (x_0, y_0, z_0) where all three species exist is

$$(x_0, y_0, z_0) = \left(\frac{\gamma A}{\gamma + \delta} + \frac{\alpha - \beta}{\gamma + \delta} \frac{\beta}{\delta}, \frac{\beta}{\delta}, \frac{\gamma}{\gamma + \delta} \left(A - \frac{\alpha}{\gamma} - \frac{\beta}{\delta} \right) \right)$$

provided

$$\gamma A > \frac{\beta - \alpha}{\gamma + \delta}, \quad A > \frac{\alpha}{\gamma} + \frac{\beta}{\delta}. \tag{5.1.1}$$

We observe that whereas α and β are predetermined by the physiology of the planktons, the rates γ and δ depend on the consumption efficiency of the phytoplankton and zooplankton. If γ and δ are large enough so that (5.1.1) holds, then (x_0, y_0, z_0) exists as a steady state. If the nutrient mass increases, the population of the zooplankton will also increase, but the lower species in the food chain, namely the phytoplankton population, will not benefit from the increase in A, i.e., y_0 is independent of A.

PROBLEM 5.9. Show that the steady state (x_0, y_0, z_0) is unstable if $k^2 < 1 + \beta k$, i.e., if $0 < k < \beta/2 + \sqrt{1 + \beta^2/4}$ and A is sufficiently large, where $k = \gamma/\delta$.

Note that k is the quotient of the rate by which phytoplankton consumes nutrients to the rate by which zooplankton consumes phytoplankton, while β is the rate by which the zooplankton returns nutrients to the environment.

Problem 5.9 asserts, in particular, the following: a situation whereby the zooplankton consumes, relatively, too much phytoplankton and recycles too little nutrients cannot sustain a large population of zooplankton.

5.2. Eigenvalue Solvers

Given a square matrix A, an eigenvalue λ and its associated eigenvector $\mathbf{x}$ $(\mathbf{x} \neq 0)$ are, by definition, a pair satisfying the equation

$$A\mathbf{x} = \lambda \mathbf{x}, \tag{5.2.1}$$

or, equivalently,

$$(A - \lambda I)\mathbf{x} = 0$$

where I is the identity matrix. This implies that

$$\det (A - \lambda I) = 0$$

which is a polynomial in λ, the *characteristic polynomial* of A. The formulas for roots of linear and quadratic equations are well known and commonly used. Even though there exist formulas for finding roots of cubic and quartic (fourth-order) equations, they are so complicated that they are rarely used.

It had been proved that there is no general formula for the roots of a fifth- or higher-order polynomials. Here we introduce two methods to compute the eigenvalues and eigenvectors of a matrix.

5.2.1. Power Method. An $n \times n$ matrix is said to be diagonalizable if it has n linearly independent eigenvectors, say $x_1, x_2, \ldots, x_n$; denote the corresponding eigenvalues by $\lambda_1, \lambda_2, \ldots, \lambda_n$. If $|\lambda_j| > |\lambda_i|$ for all $i \neq j$, then we call λ_j the **dominant eigenvalue** and x_j the **dominant eigenvector**. The simplest way to find the dominant eigenvalue is the **power method** [19]. It is based on the construction of an m-dimensional **Krylov subspace** of the form

$$\mathcal{K}_m(A, x_0) = \{x_0, Ax_0, A^2 x_0, \ldots, A^{m-1} x_0\}$$

where x_0 is a given vector.

THEOREM 5.1. *If A is an $n \times n$ diagonalizable matrix with a dominant eigenvalue, then there exists a nonzero vector x_0 such that the sequence of vectors given by $x_0, Ax_0, A^2 x_0, \ldots, A^k x_0, \ldots$ approaches a multiple of the dominant eigenvector of A.*

PROOF. We can arrange the eigenvectors $x_1, \ldots, x_n$ such that the corresponding eigenvalues satisfy

$$|\lambda_1| > |\lambda_2| \geq \ldots \geq |\lambda_n|;$$

thus λ_1 is the dominant eigenvalue. Because the n eigenvectors $x_1, x_2, \ldots, x_n$ are linearly independent, they form a basis for R^n. Clearly, a nonzero vector x_0 exists such that the linear combination

$$x_0 = c_1 x_1 + c_2 x_2 + \ldots + c_n x_n \qquad (5.2.2)$$

has nonzero leading coefficients, i.e., $c_1 \neq 0$. Applying A to (5.2.2) gives

$$\begin{aligned} Ax_0 &= c_1 Ax_1 + c_2 Ax_2 + \ldots + c_n Ax_n \\ &= c_1 \lambda_1 x_1 + c_2 \lambda_2 x_2 + \ldots + c_n \lambda_n x_n. \end{aligned}$$

Repeating this process $k - 1$ times, we get

$$\begin{aligned} A^k x_0 &= c_1 \lambda_1^k x_1 + c_2 \lambda_2^k x_2 + \ldots + c_n \lambda_n^k x_n \\ &= c_1 \lambda_1^k \left(x_1 + \frac{c_2}{c_1} \left(\frac{\lambda_2}{\lambda_1} \right)^k x_2 + \ldots + \frac{c_n}{c_1} \left(\frac{\lambda_n}{\lambda_1} \right)^k x_n \right). \end{aligned}$$

Thus

$$\lim_{k \to \infty} \lambda_1^{-k} A^k x_0 = c_1 x_1.$$

Hence $A^k x_0$ approaches a multiple of the dominant eigenvector. The convergence is geometric, with ratio λ_2/λ_1. Thus, the method converges slowly if there is an eigenvalue close in magnitude to the dominant eigenvalue.   □

After the dominant eigenvector is computed, the eigenvalue can then be obtained by the Rayleigh quotient

$$\lambda_1 = \frac{x^T \cdot Ax}{x^T \cdot x},$$

where x^T is the transpose of the column vector x. Even though the power method is simple, it computes only the dominant eigenvalue and eigenvector.

5.2.2. Shifted Inversed Power Method. Applying the power method to $(A - \mu I)^{-1}$ for a given μ, it will converge to the eigenvector corresponding to the dominant eigenvalue of the matrix $(A - \mu I)^{-1}$. The eigenvalues of this matrix are $\frac{1}{\lambda_1 - \mu}, \frac{1}{\lambda_2 - \mu}, \ldots, \frac{1}{\lambda_n - \mu}$. The largest of these numbers correspond to the smallest of $\lambda_1 - \mu, \lambda_2 - \mu, \ldots, \lambda_n - \mu$. Thus the power method will find the eigenvector whose corresponding eigenvalue is closest to μ. In the special case with $\mu = 0$, the method is applied to A^{-1} which gives the eigenvector with the smallest eigenvalue.

5.2.3. MATLAB Built-In Eigenvalue Solvers. MATLAB provides both numerical and symbolic ways to find eigenvalues of a given matrix. With one output argument, $E = \text{eig}(A)$ is a (symbolic) vector containing the eigenvalues of a square (symbolic) matrix A. With two output arguments, $[V, D] = \text{eig}(A)$ returns a matrix V whose columns are eigenvectors and a diagonal matrix D containing eigenvalues. If the resulting V is the same size as A, then A has a full set of linearly independent eigenvectors which satisfy

$$AV = VD.$$

To compute symbolic eigenvalues and eigenvectors, we need to indicate that the objects are **symbolic objects**. In the following codes, we first indicate that a, b, c, and d are symbolic objects and then build a matrix A by stacking four variables in the 2-by-2 matrix form. The function $[V,D] = \text{eig}(A)$ recognizes A as symbolic matrix so the output eigenvectors and eigenvalues are symbolic too:

```
>> syms a b c d
>> A = [a b; c d]
A = [ a, b]
   [ c, d]
>> [V,D] = eig(A)
V = [ (a/2 + d/2 - (a^2 - 2*a*d + d^2 + 4*b*c)^(1/2)/2)/c - d/c, (a/2
+ d/2 + (a^2 - 2*a*d + d^2 + 4*b*c)^(1/2)/2)/c - d/c]
   [ 1, 1]
D = [ a/2 + d/2 - (a^2 - 2*a*d + d^2 + 4*b*c)^(1/2)/2, 0] [ 0, a/2 +
d/2 + (a^2 - 2*a*d + d^2 + 4*b*c)^(1/2)/2]
```

To compute the eigenvalues of a symbolic matrix

$$A = \begin{bmatrix} a & b & 0 \\ b & a & 0 \\ 0 & 0 & 1 \end{bmatrix},$$

enter

```
>> syms a b; A = [a b 0;b a 0;0 0 1]; [V,D]=eig(A)
```
which return

V =
[0, 1, -1]
[0, 1, 1]
[1, 0, 0]
D =
[1, 0, 0]
[0, a + b, 0]
[0, 0, a - b]

However, if we replace the arguments by numerical values, the eigenvectors and eigenvalues are numerical values:

>> A = [2 1; 1 2];
>> [V,D] = eig(A)
V =
-0.7071 0.7071
0.7071 0.7071
D =
1 0
0 3

Notice that the symbolic results have bracket "[]" in the solutions.
When the size of the matrix is big, the MATLAB subroutine

eigs(A,k,sigma)

computes k eigenvalues closest to a given scalar (real or complex) sigma. See MATLAB instruction manual for details.

PROBLEM 5.10. (a) Use MATLAB symbolic calculation to get four steady states of Lotka–Volterra system (5.0.6). (b) Find the Jacobian matrix of the system (5.0.6) and then compute the eigenvalues symbolically at the biologically relevant steady state. (c) Repeat the calculation numerically in MATLAB with $r_1 = b_1 = r_2 = b_2 = 1$, $k_1 = 3$, and $k_2 = 2$ for part (b).

CHAPTER 6

Bifurcation Theory

Bifurcation theory is concerned with the question of how the behavior of a system which depends on a parameter p changes with the parameter. It focuses on any critical value, $p = p_{cr}$, where the behavior of the system undergoes radical change; such values are called **bifurcation points**. Consider for example the case of a cell which grows and replicates. During the cell cycle, there occur several critical times when the cell moves from one phase to another. The first such time occurs at the check point R_1 when the cell commits to begin replicating its DNA. The biological process about R_1 can be described by a network of signaling proteins whose dynamics is triggered by a cyclin-dependent kinase, Cdk. When the concentration p of this kinase exceeds some critical level p_{cr}, the protein network allows the cell to pass the check point R_1 and to start replicating its DNA.

Many biological processes exhibit periodic oscillations when stimulated at threshold level. This occurs, for example, in neuronal oscillations, in calcium waves that activate muscle contraction, and in cardiac rhythms. The level of simulation can be viewed as a bifurcation parameter. Mathematical theory was developed in order to explain the phenomenon associated with such periodic oscillations. The system has a stable equilibrium if the bifurcation parameter p is smaller (or larger) than a critical value p_{cr}, and as p crosses p_{cr} it gives rise to a family of periodic solutions; this type of bifurcation is called **Hopf bifurcation**, and in this chapter we explain what conditions give rise to Hopf bifurcation.

Biological switching from stability to instability, or vice versa, are quite common in biology, and they are sometimes associated with a hysteresis phenomenon. This is the case when a system can be in two states: high activation with parameter $p < p_2$ and low activation with parameter $p > p_1$, where $p_1 < p_2$. As p, at high activation, increases and crosses the higher bifurcation point p_2 the system changes from high activation to low activation. If p is then decreased, the system remains in low activation until p crosses the lower critical value p_1, at which time the system bounces back to the high activation state. Thus for $p_1 < p < p_2$ the system can be in either state, depending on its recent history. Mathematical models of hysteresis may help understand how more complex systems with hysteretic components work.

© Springer International Publishing Switzerland 2014
A. Friedman, C.-Y. Kao, *Mathematical Modeling of Biological Processes*, Lecture Notes on Mathematical Modelling in the Life Sciences, DOI 10.1007/978-3-319-08314-8_6

We consider bifurcation theory for a system of differential equations with bifurcation parameter p:

$$\frac{d\mathbf{x}}{dt} = \mathbf{f}(\mathbf{x}, p). \tag{6.0.1}$$

Bifurcation point can arise in different ways. For example, suppose a steady state of Eq. (6.0.1), which depends on p, is stable for $p < p_c$ but loses stability at p_c. Then a qualitative change has occurred in the phase portrait of the system (6.0.1), and $p = p_c$ is a bifurcation point.

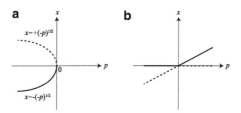

FIGURE 6.1. (**a**) Saddle-point bifurcation diagram; (**b**) transcritical bifurcation diagram

Surprisingly enough, there are just four major types of bifurcations; the first three already occur in one-dimensional systems, while the fourth needs at least two dimensions. The first three types of bifurcations and their differential equation representatives are as follows:

$$\frac{dx}{dt} = p + x^2 \quad \text{(saddle-point bifurcation)}, \tag{6.0.2}$$

$$\frac{dx}{dt} = px - x^2 \quad \text{(transcritical bifurcation)}, \tag{6.0.3}$$

$$\frac{dx}{dt} = px - x^3 \quad \text{(pitchfork bifurcation)}. \tag{6.0.4}$$

In the case of Eq. (6.0.2), there are two steady states for $p < 0$, namely $x^s_\pm = \pm\sqrt{-p}$; x^s_- is stable and x^s_+ is unstable. The solid curve in Fig. 6.1a describes the stable branch of steady states, and the dotted curve describes the branch of unstable steady states. At $p = 0$, the stable and unstable branches of steady states coalesce; the bifurcation point $p = 0$ is called a **saddle-point bifurcation**.

In the case of Eq. (6.0.3), $x = 0$ is a steady state for all p; it is stable if $p < 0$ and unstable if $p > 0$. Another branch of steady states is given by $x^s = p$; a steady state in this branch is stable for $p > 0$ and unstable for $p < 0$. This **transcritical bifurcation** diagram is shown in Fig. 6.1b; the diagram is characterized by an exchange of stability of the branches of steady states at the bifurcation point $(p = 0)$.

In the case of Eq. (6.0.4), for $p < 0$ the only steady state is $x^s = 0$; but for $p > 0$, there are two more steady states, namely, $x^s_\pm = \pm\sqrt{p}$. The steady state $x^s = 0$ is stable if $p < 0$ and unstable if $p > 0$. The steady states $x_\pm^s = \pm\sqrt{p}$ (for $p > 0$) are both stable. The diagram of this **pitchfork bifurcation** is shown in Fig. 6.2a. A similar pitchfork bifurcation for the equation

$$\frac{dx}{dt} = px + x^3 \tag{6.0.5}$$

is shown in Fig. 6.2b; in this case the two branches $x_\pm^s = \pm\sqrt{-p}$ are unstable, and $x^s = 0$ is stable if $p < 0$ and unstable if $p > 0$. The bifurcation in Fig. 6.2a is said to be **supercritical** since the bifurcation branches appear for values of p larger than the bifurcation point $p_c = 0$. The bifurcation in Fig. 6.2b is called **subcritical** since the two bifurcating branches occur for p smaller than the bifurcation point $p_c = 0$.

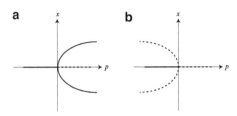

FIGURE 6.2. Pitchfork bifurcation. *Solid curves* represent stable steady states, while *dotted curves* are unstable steady states

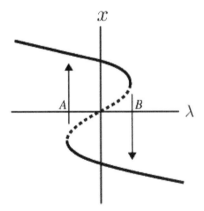

FIGURE 6.3. Steady-state bifurcation diagram showing bistability in the range $\alpha_0 < \lambda < \alpha_1$. The *arrows* indicate components of the hysteresis loop

A bifurcation diagram distinguishes one of the parameters of the system as the bifurcation parameter. Consider the equation

$$\frac{dx}{dt} = -x^3 + \mu x - \lambda \quad (\mu > 0) \tag{6.0.6}$$

and choose λ as the bifurcation parameter. One can show that for λ in a range $\alpha_0 < \lambda < \alpha_1$ there exist three steady states, while for $\lambda < \alpha_0$ or $\lambda > \alpha_1$ there exists only one steady state. Figure 6.3 shows the bifurcation diagram: points on the solid curves are stable steady points; the middle branch corresponds to unstable steady states. The bifurcation points $\lambda = \alpha_0$ and $\lambda = \alpha_1$ are saddle-point bifurcations.

If, in steady state on the upper branch in Fig. 6.3, the bifurcation parameter λ is increased (slowly enough for the system to settle to steady states), the value α_1 will be reached where a discontinuous drop in the steady state will occur, as indicated by the vector B. If the process is repeated, but this time starting from values larger than α_1 and slowly decreasing λ, the state α_0 will be reached and the steady state will jump from the lower branch to the upper branch, as indicated by the vector A. The loop formed by these two discontinuous transitions is referred to as a **hysteresis loop**. Thus, for values of λ within the bistable range $\alpha_0 < \lambda < \alpha_1$, which of the two stable steady states of the system is reached depends on the initial condition.

PROBLEM 6.1. Draw, by hand, the bifurcation diagram for the following equation, with p as the bifurcation parameter:

$$\frac{dx}{dt} = px - x^2 - x^3.$$

Find the bifurcation points and classify them.

Many biological processes are oscillatory in nature, for example, the beating of the heart, spiking of neurons in the brain, circadian rhythms arising from cycles of day and night, and the cell-division cycle. The existence and characteristics of these oscillations depend on the parameters of the system, and it is important to know when or how these oscillations arise.

The **Hopf bifurcation** refers to the bifurcation of periodic solutions from a steady-state solution as a parameter is varied.

Consider the system

$$\frac{dx}{dt} = f(x, y, p), \qquad \frac{dy}{dt} = g(x, y, p). \tag{6.0.7}$$

Assume that for all p in some interval there exists a steady state $(x^s(p), y^s(p))$ and that the two eigenvalues of the Jacobian matrix (evaluated at the steady state) are complex numbers $\lambda_1(p) = \alpha(p) + i\beta(p)$ and $\lambda_2(p) = \alpha(p) - i\beta(p)$. Assume also that for some parameter p_0 within the interval the following are true:

$$\alpha(p_0) = 0, \quad \beta(p_0) \neq 0, \quad \text{and} \quad \frac{d\alpha}{dp}(p_0) \neq 0.$$

THEOREM 6.1. *Under the above conditions one of the following three cases must occur:*

(1) *there is an interval $p_0 < p < c_1$ such that for any p in this interval there exists a unique periodic orbit containing $(x^s(p_0), y^s(p_0))$ in its interior and having a diameter proportional to $|p - p_0|^{1/2}$;*

(2) *there is an interval $c_2 < p < p_0$ such that for any p in this interval there exists a unique periodic orbit as in case (1);*

(3) *for $p = p_0$ there exist infinitely many orbits surrounding $(x^s(p_0), y^s(p_0))$ with diameters decreasing to zero.*

A proof of Theorem 6.1 is given, for instance, in [14].

Case (1) is called a **supercritical Hopf bifurcation** and case (2) is called a **subcritical Hopf bifurcation**; these cases are generic. Case (3) is rather infrequent. Figure 6.4 illustrates a supercritical Hopf bifurcation.

We illustrate the statement of Theorem 6.1 in a special case where all the steady points $(x^s(p), y^s(p))$ coincide with the origin. The system we consider is given by

$$\frac{dx_1}{dt} = px_1 - \mu x_2 - ax_1 \left(x_1^2 + x_2^2\right),$$
$$\frac{dx_2}{dt} = \mu x_1 + px_2 - ax_2 \left(x_1^2 + x_2^2\right),$$

(6.0.8)

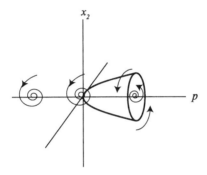

FIGURE 6.4. Supercritical Hopf bifurcation at $p = 0$

where μ, a are positive constants. The point $\mathbf{x} = 0$ is a steady state for all p. The eigenvalues $\lambda(p) = \alpha(p) \pm i\beta(p)$ of the Jacobian matrix at $\mathbf{x} = 0$ are given by $\lambda = p \pm i\mu$, so that $\alpha(p) = p$, $\beta(p) = \mu$ and $\alpha(0) = 0$, $\beta(0) = \mu > 0$, and $d\alpha(0)/dp = 1 > 0$. Thus the Hopf bifurcation occurs at $p = 0$. The steady point $\mathbf{x} = 0$ is stable if $p < 0$ and unstable if $p > 0$. For any $p > 0$ there exists a periodic solution

$$x_1(t) = \sqrt{\frac{p}{a}} \cos \mu t, \quad x_2(t) = \sqrt{\frac{p}{a}} \sin \mu t,$$

with period $2\pi/\mu$. The trajectory is a circle of radius $\sqrt{p/a}$, i.e., it is proportional to $p^{1/2}$, as asserted in Theorem 6.1. The Hopf bifurcation is supercritical, since the periodic solutions occur for $p > 0$.

So far we have assumed that $a > 0$. The case $a < 0$ can be analyzed in a similar way. Consider next the case $a = 0$. Then the eigenvalues of the linear system at $p = 0$ are $\pm i\mu$. This is the case of a center, shown in Fig. 3.1f; for $p = 0$ there are infinitely many periodic solutions, as asserted in case (3) of Theorem 6.1.

PROBLEM 6.2. Consider a solution of (6.0.8) with initial condition such that $x_1^2(0) + x_2^2(0) > \frac{p}{a}$. Set $R(t) = x_1^2(t) + x_2^2(t)$ and show that

$$\frac{dR}{dt} = pR - aR^2.$$

Deduce that $R(t) > \frac{p}{a}$ for all $t > 0$,

$$\frac{1}{2p} \ln \left(\frac{R(t)}{R(t) - p/a} \right) = t + const,$$

and $R(t) \to \frac{p}{a}$ as $t \to \infty$.

The last conclusion of Problem 6.2 shows that the periodic solution with radius $\sqrt{p/a}$ is stable from the outside. The next problem shows that the periodic solution is not stable from the inside.

PROBLEM 6.3. If in Problem 6.2 we take

$$x_1^2(0) + x_2^2(0) < \frac{p}{a},$$

then $x_1^2(t) + x_2^2(t) \to 0$ as $t \to \infty$.

PROBLEM 6.4. Consider the system

$$\begin{aligned}
\frac{dx}{dt} &= x^2 (1 - x) - xy, \\
\frac{dy}{dt} &= 4xy - 4py,
\end{aligned} \tag{6.0.9}$$

where $0 < p < 1$. Show that it has an equilibrium point $(p, p(1 - p))$ and that Hopf bifurcation occurs at $p = \frac{1}{2}$. The above system may be viewed as a predator–prey model with x being the prey.

The Hopf bifurcation occurs in many biological processes. We give here another example of a predator–prey model, where the prey u has a logistic growth:

$$\begin{aligned}
\frac{du}{dt} &= u\left(1 - \tfrac{1}{2}u\right) - uv, \\
\frac{dv}{dt} &= -v(1 - pv) + 2uv,
\end{aligned} \tag{6.0.10}$$

where $0 < p < 1$. The term pv accounts for the fact that as v decreases (by death) there will be more resources available for the live predators, resulting in increased longevity.

PROBLEM 6.5. Show that in the model (6.0.10)

$$(u, v) = \left(2 - \frac{6}{4 - p}, \frac{3}{1 - p}\right)$$

is a steady point which is stable if $0 < p < \frac{1}{4}$ and unstable if $\frac{1}{4} < p < 1$ and that Hopf bifurcation occurs at $p = \frac{1}{4}$.

6.1. Numerical Computation of Bifurcation Diagrams

In MATLAB there is no command to generate bifurcation diagram directly. However, based on the definition of the bifurcation diagram, we can draw the zeros of (6.0.1) and then identify stable and unstable branches. For example, consider the saddle-point bifurcation diagram for

$$p + x^2 = 0,$$

and create a MATLAB function called saddlefun:

    function y=saddlefun(x,p)
    y=p+x.^2;

The bifur.m in Algorithm 2 computes the bifurcation diagram. The inputs are (fcn, xrange, prange) which specify the function and the range of x and p. It first uniformly samples 101×101 points (x_i, p_j) in the range of x and p by using meshgrid and then evaluates $f(x_i, p_j)$. The zero contours shown in Fig. 6.5a are drawn via the built-in function contour. We then select the points shown in green color in Fig. 6.5b which has $|f(x_i, p_j)| > 0.1\text{mean}(|f(x_i, p_j)|)$. This selection process can avoid selecting the points on the nullcline. If we solve the ODE

$$\frac{dx}{dt} = p + x^2,$$

with these initial conditions by using Euler method with step size as 0.1, the points will either get away from the unstable branches or come close to the stable branches. We then draw the solution at the later time as blue dots as shown in Fig. 6.5c.

In Fig. 6.6, we draw the solution of the ODE system (6.0.8) which has Hopf bifurcation behavior. The parameters are $\mu = a = 1$. We compute the ODE system by MATLAB built-in function "ode23" [1] with 16 initial conditions:

$$(x_1^{i1}, x_2^{i1}) = 2\left(\cos\left(\frac{2\pi}{8}(i - 1)\right), \sin\left(\frac{2\pi}{8}(i - 1)\right)\right), \quad i = 1 : 8,$$

$$(x_1^{i2}, x_2^{i2}) = 0.01\left(\cos\left(\frac{2\pi}{8}(i - 1)\right), \sin\left(\frac{2\pi}{8}(i - 1)\right)\right), \quad i = 1 : 8.$$

The solutions are plotted for p in the range of $[-2, 2]$ with the increment 0.5. We indicate the trajectories with $\left\|(x_1^{i1(i2)}, x_2^{i1(i2)})\right\| = 2$ (0.01) in blue (red) color. We can see that the point $\mathbf{x} = 0$ is a steady state, stable if $p < 0$ and

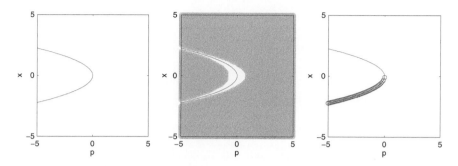

FIGURE 6.5. The MATLAB plot of saddle-point bifurcation:
(a) $p + x^2 = 0$, (b) initial points in *green*, (c) *blue points*
which converge to stable branches

unstable if $p > 0$. For any $p > 0$ there is a periodic solution. The codes that
generate Fig. 6.6 are in Algorithms 3 and 4.

PROBLEM 6.6. Use MATLAB to generate the transcritical bifurcation
and pitchfork bifurcation.

PROBLEM 6.7. Use MATLAB to generate the steady-state bifurcation
diagram showing bistability as shown in Fig. 6.3.

Algorithm 2 bifur.m

```
% BIFUR Draws bifurcation diagrams
% BIFUR(FCN,XRANGE,PRANGE) draws the bifurcation diagram
% for the function FCN over the specified x and p ranges.
% FCN is a handle to a user-defined function that takes as
% arguments a variable x and a parameter p. XRANGE is a
% row vector of the form [XMIN XMAX]. PRANGE is a row vector
% of the form [PMIN PMAX].
%
% Example:
% bifur(@saddlefun,[-5 5],[-5 5]);
%
% where saddlefun is a user-defined function of the form
%
% function y=saddlefun(x,p)
% y=p+x.^2;
%
function bifur(fcn,xrange,prange)
nn = 100; % number of points plotted in each range
p1 = [prange(1):(prange(2)-prange(1))/nn:prange(2)]; % sample points in p
x1 = [xrange(1):(xrange(2)-xrange(1))/nn:xrange(2)]; % sample points in x
[p,x] = meshgrid(p1,x1); % generate grid points in [p,x]
fval = feval(fcn,x,p); % evaluate the points value
figure(1);[c,h] = contour(p,x,fval,[0,0],'r'); % plot the zero contour line
xlabel('p') ylabel('x')
x = x(:); p = p(:);
ind = find(abs(fval)>0.1*mean(abs(fval(:))));
x = x(ind); p = p(ind);
figure(1); hold on; plot(p,x,'go') % draw the initial points
for iter = 1:100
x = x+ 0.1*feval(fcn,x,p); % solve ode by Euler method with step size
equals 0.1
end
hold on; plot(p(:),x(:),'bo')
```

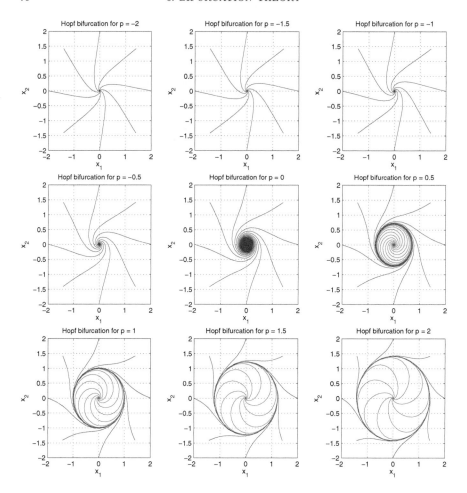

FIGURE 6.6. The phase plane varies with respect to p for (6.0.8) with $m = a = 1$

Algorithm 3 main_hopf.m

```
% Example of Supercritical Hopf bifurcation
% From a stable node to a stable limit cycle shown in the phase-plane.
% Phase-plane plots of the components [x_1,x_2] from p= -2 to p=2.
global p
for p = -2:0.5:2;
tfinal=1000;
% First set of initial values
pts = 8;
theta = 0:2*pi/pts:2*pi-2*pi/pts;
x1 = 2*[cos(theta); sin(theta)];
% Second set of initial values
x2 = 0.01*[cos(theta); sin(theta)];
figure(1);clf; hold on;
for i = 1:pts
[t x1t]=ode23(@hopf,[0 tfinal],x1(:,i));
plot(x1t(:,1),x1t(:,2),'b')
[t x2t]=ode23(@hopf,[0 tfinal],x2(:,i));
plot(x2t(:,1),x2t(:,2),'r')
end
title(['Hopf bifurcation for p = ', num2str(p)]); grid on; xlabel('x_1');
ylabel('x_2'); box on;
pause(1) % pause for one second so that the figure can be seen
end
```

Algorithm 4 hopf.m

```
function xdot=hopf(t,x)
global p
mu = 1; a = 1; xdot = zeros(2,1);
xdot(1)=-mu*x(2)+x(1)*(p-a*(x(1)^2)-a*(x(2)^2));
xdot(2)=mu*x(1)+x(2)*(p-a*(x(1)^2)-a*(x(2)^2));
```

CHAPTER 7

Neuronal Oscillations

A neuron is a cell that receives, conducts, and transmits signals. It receives signals as tiny electric pulses from many dendritic branches. The electric pulses arrive at one location near the cell body, called **soma**, which houses the nucleus of the nerve cell. If the combined signal exceeds a certain threshold, then it triggers initiation of a wave of electrical oscillation that travels along the membrane (also called the **plasma membrane**) of a relatively long axon, as shown schematically in Fig. 7.1. The electrical wave arrives at the terminal edge of the axon, where it transmits the signal to dendritic branches of another neuron, to muscle cells, or to glands. Neurons receive signals from the sense organs inward, and after the signals are processed in the brain, or in the spinal cord, a response by other neurons is sent onward for action by muscle cells and glands.

In this chapter we focus on the traveling electric signal along the axon; the electric wave is called the **action potential**. If the plasma membrane were an ordinary conductor, then the electric pulse of the action potential would weaken substantially upon arrival at the terminal end, and animal activities would be slowed down to near paralysis. That this is not the case is due to the special **voltage-gated channels** distributed along the plasma membrane, and to unequal concentrations of ions species inside and outside the neuron, as we proceed to explain.

Within every cell, as well as on the outside, there are sodium ions Na^+ and potassium ions K^+. The concentration of sodium is larger outside than inside the cell and the concentration of potassium is larger inside than outside the cell. Hence when a sodium channel on the plasma membrane of the axon opens, sodium ions flow into the cell, thereby increasing the electric potential (i.e., the positive charge) inside the axon. Similarly, when a potassium channel opens, potassium ions flow outside the axon, thereby decreasing the electric potential inside the axon.

The difference between the electric potential inside and outside the plasma membrane is usually called the **voltage** of the axon. When a neuron is at rest, the electric potential inside the plasma membrane is negative compared to the electric potential outside the membrane. The difference between these two potentials is called the **resting potential**; it is negative. The sodium

© Springer International Publishing Switzerland 2014
A. Friedman, C.-Y. Kao, *Mathematical Modeling of Biological Processes*, Lecture Notes on Mathematical Modelling in the Life Sciences, DOI 10.1007/978-3-319-08314-8_7

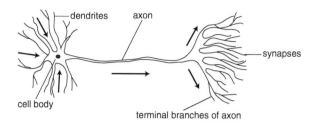

FIGURE 7.1. A neuron. The *arrows* indicate direction of signal conduction

and potassium gates mentioned above open or close in response to the level of the voltage; this accounts for their name, voltage-gated channels.

When positive charge enters the neuron through the dendritic branches, if the combined charge exceeds a threshold level, then the neuron fires: the sodium channels begin to open, sodium ions flow into the axon, and electric potential is increased. This positive feedback continues to increase the voltage until the voltage reaches a high enough level at which time the open sodium channels begin to slowly shut down and the voltage begins to drop. A similar role is played by the potassium; but potassium channels open a little bit after the sodium channels.

Hodgkin and Huxley measured the action potential along axons and in the early 1950s developed a mathematical model whose predictions fit exceedingly well with their measurements. They were awarded the Nobel prize for this work.

Subsequently, FitzHugh and Nagumo introduced a simpler mathematical model with only one type of channels. Their model is capable of producing qualitatively the traveling wave behavior of the action potential. Here we shall describe the FitzHugh–Nagumo model based only on voltage-gated sodium channels. But first we need to developing some mathematical theory on singular perturbations in dynamical systems.

Biological processes often involve several time scales. Consider the following system with two time scales:

$$\frac{dx}{dt} = f(x, y), \qquad (7.0.1)$$

$$\frac{dy}{dt} = \varepsilon g(x, y), \qquad (7.0.2)$$

where ε is a small positive parameter; the variable y changes at a smaller rate than the variable x. Assume that the x-**nullcline** is cubic and the y-**nullcline** is a monotone function and that these nullclines intersect at one point Q, as shown in Fig. 7.2. Then, there exists a periodic orbit of Eqs. (7.0.1) and (7.0.2) which lies in the vicinity of the curve ABCDA. In order to explain why this is so, observe that if $\varepsilon = 0$, then $dy/dt = 0$ so that $y(t) = \text{constant}$. Taking $\varepsilon = 0$ as an approximation is justified in case $f(x, y)$ stays away from zero in Eq. (7.0.1), for then dy/dt is relatively negligible compared to dx/dt. Thus, a trajectory $(x(t), y(t))$ starting at (x_0, y_0) where $f(x_0, y_0) \neq 0$ will approximately satisfy $y(t) = y_0$ as long as $f(x(t), y_0)$ remains not equal to zero. In particular, a trajectory starting near and below B will travel horizontally, with speed $dx/dt = f > 0$, until it reaches the $x-$nullcline near C. Similarly, a trajectory starting near and above D will travel horizontally, with $dx/dt = f < 0$, until it reaches the $x-$ nullcline near A.

To see how a trajectory evolves once it is located on the $x-$ nullcline, one introduces the slow time scale τ defined as $\tau = \varepsilon t$. Then Eqs. (7.0.1) and (7.0.2) become

$$\varepsilon \frac{dx}{d\tau} = f(x, y), \quad \frac{dy}{d\tau} = g(x, y). \tag{7.0.3}$$

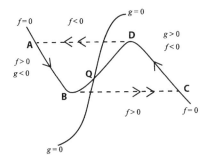

FIGURE 7.2. Singular perturbation analysis of an oscillatory system. The *curves* corresponding to $f = 0$ and $g = 0$ are the nullclines of the system in Eqs. (7.0.1) and (7.0.2)

Since $g < 0$ above the $y-$ nullcline, g is negative along the arc AB and thus it pulls y further downward toward B as well as at B. Since B is the point at which the $x-$ nullcline turns upward, the trajectory then departs from $x-$ nullcline, and then as explained above, it proceeds along a curve close to the horizontal line BC. In a similar way, using the slow time scale, one can see that the trajectory will evolve approximately along the $x-$ nullcline from C to D (since $dy/d\tau = g > 0$ on CD) and then back to A along $y = const$. The motion along the $x-$ nullcline takes place in the slow time scale (slow motion), whereas the horizontal motion takes place on the original (fast) time scale as indicated by the double arrows in Fig. 7.2.

In the above discussion we have tacitly assumed that a trajectory $(x(t), y(t))$ starting at A will remain on the arc AB of the $x-$nullcline, or at least very close to it, as $y(t)$ changes. This is indeed the case because $\partial f(x, y)/\partial x < 0$ along the arc AB and hence (x, y) is a stable equilibrium point for the differential equation $dx/d\tau = \frac{1}{\varepsilon} f(x, y)$, with y fixed.

The above considerations can be made more precise. One can prove mathematically that for any small $\varepsilon > 0$ there exists a periodic orbit near the curve $ABCDA$. The above example is generic; it can be extended to the case of more than two ODEs showing, for example, the existence of periodic oscillations in networks.

PROBLEM 7.1. Consider the system

$$\frac{dx}{dt} = x + y, \quad \frac{dy}{dt} = \epsilon(x + y)^2,$$

with $x(0) = 0, y(0) = 1$. Show that the solution exists only for a bounded interval $0 \leq t < T_\epsilon$ and compute T_ϵ.

PROBLEM 7.2. Consider the system

$$\frac{dx}{dt} = x(x - 1)(2 - x) - y, \quad \frac{dy}{dt} = \epsilon(\lambda x - y - 100).$$

Compute, by hand, the values of λ for which a periodic solution exists when ϵ is sufficiently small, i.e., for which the phase portrait is as in Fig. 7.2.

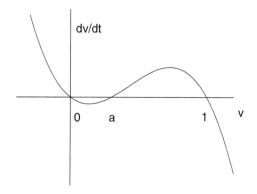

FIGURE 7.3. Profile of dv/dt as a function of v

We now return to the FitzHugh–Nagumo model and denote by v the voltage of the action potential. By scaling the voltage v, we set

$v = 0,$ the resting potential,
$v = a,$ the threshold above which the
           neuron fires,
$v = 1,$ the voltage level at which the
           sodium channels have completely closed.

We are going to develop a differential equation for $v = v(t)$. To do that we need to describe dv/dt as a function of v. Since $v = 0$ is a stable point (small signals do not make the neuron fire), $dv(0)/dt < 0$. When the sodium channels begin to open the voltage increases, so $dv/dt > 0$ and v continues to increase as the neuron fires at $v = a$, hence $dv(a)/dt > 0$. Eventually the voltage v decreases until the sodium channels close at $x = 1$, so that $dv(1)/dt < 0$. The simplest expression for dv/dt as a function $f(v)$ of v is illustrated in Fig. 7.3.

A formula that is consistent with the profile of Fig. 7.3 is given by

$$\frac{dv}{dt} = -v(v - a)(v - 1) \equiv f(v) \tag{7.0.4}$$

What is missing in (7.0.4), however, is the part of the process which accounts for closing of the sodium channels. We introduce a channel blocking mechanism by a variable w which acts to diminish v, replacing (7.0.4) by

$$\frac{dv}{dt} = -v(v - a)(v - 1) - w \equiv f(v) - w. \tag{7.0.5}$$

We assume that dw/dt increases linearly in v and that w degrades linearly:

$$\frac{dw}{dt} = \varepsilon(v - \gamma w) \tag{7.0.6}$$

Here γ and ε are positive constants, but ε is very small since the channels open and close very slowly.

So far we have not incorporated into the model the applied current I coming from the soma. We incorporate it by adding I to the right-hand side of (7.0.5):

$$\frac{dv}{dt} = -v(v - a)(v - 1) - w + I \equiv f(v) - w + I. \tag{7.0.7}$$

It is easily seen, by drawing, that for any a, $0 < a < 1$, there is a $\gamma = \gamma(a)$ such that the $v -$ nullcline of (7.0.5) and $w -$ nullcline of (7.0.6) intersect at one point if $\gamma < \gamma(a)$ and at three points if $\gamma > \gamma(a)$.

PROBLEM 7.3. Let $f(v_1) = \min_{0 \leq v \leq 1} f(v)$, $f(v_2) = \max_{0 \leq v \leq 1} f(v)$ and denote by Γ the curve $\{f(v), v_1 < v < v_2\}$. Show that $f'(v) > 0$ on Γ, and $f'(v) < 0$ if $0 \leq v < v_1$ or $v_2 < v \leq 1$.

It is easily seen, by drawing, that for any $0 < a < 1$, if γ is near 0, then the $w -$ nullcline of (7.0.6) is nearly in the vertical direction. Hence there is an interval (α, β) such that the $w -$ nullcline of (7.0.6) intersects the $v -$ nullcline of (7.0.7) at a single point which lies on Γ whenever $\alpha < I < \beta$. This situation is valid if γ is increased up to some number γ_*.

We shall now apply the singular perturbation analysis to the system (7.0.6)–(7.0.7). For any $0 < a < 1$, if γ is sufficiently small, then the $w -$ nullcline intersects the $v -$ nullcline on the arc where $v_1 < v < v_2$, provided the applied current exceeds a certain threshold, and then the action potential develops periodic oscillations provided ε is sufficiently small.

PROBLEM 7.4. Given any values of a, γ, I such that the nullclines of (7.0.6)–(7.0.7) intersect at a single point $P = (u_*, v_*)$ lying on the monotone increasing arc of the $v-$nullcline, show that P is stable if ε is sufficiently small.

This result means that neuronal oscillations cannot converge to a steady state.

7.1. Numerical Computation of Neuronal Oscillations

In Fig. 7.4a, we draw the function $f(v)$ in (7.0.4) versus v and the solution of the ODE (7.0.4) with different initial conditions from 0.1 to 1.1 with the increment 0.1 by using Algorithms (5) and (6). The numerical implementation is based on MATLAB built-in ODE solvers "ode45" discussed in Sect. 4.3.3.

Algorithm 5 main_potential.m

```
global a
a = 0.3; v = -0.15:0.01:1.1; t = 0;
dvdt = potential_eqn(t,v);
figure(1);plot(v,dvdt)
xlabel('v')
ylabel('dv / dt')
options = odeset('RelTol',1e-4,'AbsTol',[1e-4]);
t0 = 0; tfinal = 10;
v_ini = 0.1:0.1:1.1;
nn = length(v_ini);
for ind = 1:nn
y_ini = v_ini(ind);
[T,y] = ode45(@potential_eqn,[t0 tfinal],y_ini,options);
figure(2);hold on;plot(T,y); xlabel('t'); ylabel('v')
end
```

Algorithm 6 potential_eqn.m

```
function dv = potential_eqn(t,v)
global a
dv = -v.*(v-a).*(v-1);
```

Algorithms 7 and 8 solve Fitzhugh–Nagumo model with $I = 0$ and three different initial conditions $(0.4, 0)$, $(0.5, 0)$, and $(0.6, 0)$. In Fig. 7.5a, we draw these three solutions of Fitzhugh–Nagumo model in $x_1 - x_2$ plane. Notice that the nullclines intersect at one equilibrium point $(0, 0)$. We see that three trajectories converge to the equilibrium point. In Fig. 7.5b, the solutions of $(v(t), w(t))$ are drawn versus time t.

PROBLEM 7.5. Under the same setup as in Problem 7.4, compute the solution of (7.0.6)–(7.0.7) initiating at (u_0, v_0) near P (but $(u_0, v_0) \neq P$) when $\varepsilon = 1, \varepsilon = \frac{1}{10}, \varepsilon = \frac{1}{100}$.

PROBLEM 7.6. Do the same computations as in Problem 7.5 when the initial point (u_0, v_0) lies on the $v -$ nullcline.

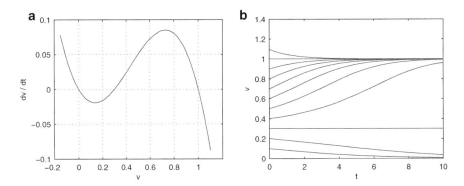

FIGURE 7.4. (a) The figure of $\frac{dv}{dt} = f(v) = -v(v-a)$ $(v-1) - w$ (7.0.4) versus v with $a = 0.3$ and $w = 0$. (b) The solution of ODE with different initial conditions $v(0)$ from 0.1 to 1.1 with the increment 0.1

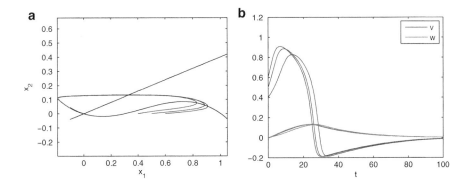

FIGURE 7.5. (a) The solutions of Fitzhugh–Nagumo model with three different initial conditions $(0.4, 0)$, $(0.5, 0)$, and $(0.6, 0)$ (a) in $x_1 - x_2$ plane. (b) The solutions of $(v(t), w(t))$ versus time t

---

Algorithm 7 main_FitzhughNagumo.m

```
% This is the main function to solve
% the Fitzhugh-Nagumo model
global a epsi gamma
a = 0.3; epsi = 0.01; gamma = 2.5;
options = odeset('RelTol',1e-4,'AbsTol',[1e-4 1e-4]);
t0 = 0; tfinal = 100;
v_ini = [0.4 0.5 0.6]; w_ini = [0.0 0.0 0.0];
for ind = 1:3
y_ini = [v_ini(ind) w_ini(ind)];
[T,y] = ode45(@FitzhughNagumo_eqn,[t0 tfinal],y_ini,options);
% plot N2 versus N1
figure(1);hold on; plot(y(:,1),y(:,2));
xlabel('x_1')
ylabel('x_2')
axis equal; box on
figure(2); hold on; plot(T,y(:,1),T,y(:,2))
legend('v','w')
end
% plot the nullcline
v = -0.1:0.01:1.05;
w1 = -v.*(v-a).*(v-1);
figure(1); hold on; plot(v,w1,'k')
w2 = v/gamma;
figure(1); hold on; plot(v,w2,'k')
```

Algorithm 8 FitzhughNagumo_eqn.m

```
function dy = FitzhughNagumo_eqn(t,y)
global a epsi gamma
dy = zeros(2,1); % need to be a column vector
dy(1) = -y(1)*(y(1)-a)*(y(1)-1)-y(2); dy(2) = epsi*(y(1)-gamma*y(2));
```

CHAPTER 8

Conservation Laws

In this chapter we develop the mathematics of conservation laws for concentration of species. Such laws are used in many biological processes. Two quite different examples will be given. The first example is about age structure of cells in connection with targeted therapeutics of cancer. The second example is concerned with the alarming spread of antibiotic resistance of organisms in hospitals.

Consider a chemical species with concentration $c(x,t)$ moving with velocity $v(x,t)$ in the positive $x -$ direction. For any small spatial interval $(x, x + \Delta x)$ and small time interval $(t, t + \Delta t)$, the change in the mass of c in the interval $(x, x + \Delta x)$ during this time interval

$$\int_x^{x+\Delta x} [c(\xi, t + \Delta t) - c(\xi, t)] \, d\xi,$$

is equal to the total mass that entered or left the interval at x minus the total mass that entered or left at $x + \Delta x$:

$$\int_t^{t+\Delta t} [v(x, s)c(x, s) - v(x + \Delta x, s)c(x + \Delta x, s)] \, ds.$$

Equating the two integrals, after dividing by $\Delta x \cdot \Delta t$, and taking $\Delta x \to 0$, $\Delta t \to 0$, we obtain the **conservation of mass law**:

$$\frac{\partial c}{\partial t} + \frac{\partial}{\partial x}(vc) = 0.$$

If the concentration c grows at rate $R(c, x, t)$, then

$$\frac{\partial c}{\partial t} + \frac{\partial}{\partial x}(vc) = R(c, x, t). \tag{8.0.1}$$

Similarly, if concentration $c = c(\mathbf{x}, t)$, where $\mathbf{x} = (x_1, \ldots, x_n)$, is moving with velocity $\mathbf{v} = (v_1, \ldots, v_n)$ where $v_i = v_i(\mathbf{x}, t)$, then

$$\frac{\partial c}{\partial t} + \nabla \cdot (\mathbf{v}c) = R(c, \mathbf{x}, t).$$

© Springer International Publishing Switzerland 2014
A. Friedman, C.-Y. Kao, *Mathematical Modeling of Biological Processes*, Lecture Notes on Mathematical Modelling in the Life Sciences, DOI 10.1007/978-3-319-08314-8_8

Consider next several species with concentrations c_i $(1 \leq i \leq m)$ such that c_i is transformed into c_j at a rate R_{ij}. Then the conservation of mass law takes the form

$$\frac{\partial c_i}{\partial t} + \nabla \cdot (\mathbf{v}_i c_i) = \sum_{j=1}^{m} R_{ji} + R_i \tag{8.0.2}$$

where $R_{ii} = -\sum_{j=1, j \neq i}^{m} R_{ij}$, R_i is the growth rate in c_i, and $\mathbf{v}_i$ is the velocity of species c_i. Note that

$$\sum_{i=1}^{m} \sum_{j=1}^{m} R_{ji} = 0. \tag{8.0.3}$$

The conservation of mass law can be applied not only to concentration of chemical species, but to any concentration of species, e.g., bacteria, proteins, and people.

Consider the initial-value problem consisting of solving ρ from the equation

$$\rho_t + \mathrm{div}\,(\rho \mathbf{v}) = f, \quad \mathbf{x} \in R^n \tag{8.0.4}$$

and the initial condition

$$\rho(\mathbf{x}, 0) = \rho_0(\mathbf{x}) \tag{8.0.5}$$

where $\mathbf{v} = \mathbf{v}(\mathbf{x}, t)$, $f = f(\mathbf{x}, t)$, and R^n is the space of points $\mathbf{x} = (x_1, \ldots, x_n)$ with $-\infty < x_i < \infty$ $(i = 1, \ldots, n)$. One method of solving this problem is the **method of characteristics**. Introduce the characteristic curves that end at (x, t):

$$\frac{d\xi_i}{d\tau} = v_i(\boldsymbol{\xi}, \tau), \quad 0 < \tau < t \quad (i = 1, \ldots, n)$$

$$\xi_i(t) = x_i \quad (\text{that is, } \boldsymbol{\xi}(t) = \mathbf{x}),$$

where $\mathbf{v}(\boldsymbol{\xi}, \tau) = (v_1(\boldsymbol{\xi}, \tau), \ldots, v_n(\boldsymbol{\xi}, \tau))$, and denote the solution by $\boldsymbol{\xi}(\tau; \mathbf{x}, t)$. Set $\mathbf{x}_0 = \boldsymbol{\xi}(0; \mathbf{x}, t)$, $\bar{\rho}(\tau) = \rho(\boldsymbol{\xi}(\tau; \mathbf{x}, t), \tau)$. Then

$$\frac{d\bar{\rho}}{d\tau} = \sum_{i=1}^{n} \frac{\partial \rho}{\partial \xi_i} \frac{\partial \xi_i}{\partial \tau} + \frac{\partial \rho}{\partial \tau} = \sum_{i=1}^{n} v_i \frac{\partial \rho}{\partial \xi_i} + \frac{\partial \rho}{\partial \tau}$$

$$= \frac{\partial \rho}{\partial \tau} + \sum_{i=1}^{n} \frac{\partial}{\partial \xi_i}(\rho v_i) - \sum_{i=1}^{n} \bar{\rho} \frac{\partial v_i}{\partial \xi_i},$$

or, by Eq. (8.0.4), and the initial condition (8.0.5),

$$\frac{d\bar{\rho}}{d\tau} = -\bar{\rho}\mathrm{div}(\mathbf{v}) + f \quad \text{for} \quad 0 < \tau < t,$$
$$\bar{\rho}(0) = \rho_0(\mathbf{x}_0).$$

This is an ODE problem and its solution yields the solution to Eqs. (8.0.4)–(8.0.5):

$$\rho(\mathbf{x}, t) = \bar{\rho}(t) = \rho_0(\boldsymbol{\xi}(0; \mathbf{x}, t), 0).$$

In the case where the function $\rho(\mathbf{x}, t)$ is restricted to $\mathbf{x} \in \Omega$, where Ω is bounded domain, the characteristic curves $\boldsymbol{\xi}(\tau)$ from points $(\mathbf{x}, t)$ may reach the boundary $\partial\Omega$ at some time $\tau > 0$; in this situation, boundary values must be prescribed at these "exit points" of $\partial\Omega$. As a simple example, consider the case

$$\frac{\partial\rho}{\partial t} + \alpha\rho = 0 \quad \text{for} \quad 0 < x < A,$$

where α is a positive constant. Then, the characteristic curves are defined by

$$\frac{d\xi}{d\tau} = \alpha \quad \text{or} \quad \xi(\tau) = (\text{x-}\alpha t) + \alpha\tau \quad (\text{with } \xi(t) = x).$$

If $x > \alpha t$ then the characteristic curve does not exit the interval $(0, A)$ for all $0 < \tau < t$; but if $x < \alpha t$ then it exits this interval at $\xi = 0$ at time $\tau = t - x/\alpha$. Thus, initial and boundary values must be assigned as follows:

$$\begin{aligned} \rho(x,0) &= \rho_0 \quad &\text{for} \quad 0 < x < A, \\ \rho(0,t) &= \rho_1(t) \quad &\text{for} \quad 0 < t < \infty. \end{aligned}$$

Suppose, for any $\mathbf{x} \in \partial\Omega$, the characteristic curves $\boldsymbol{\xi}(\tau; \mathbf{x}, t)$ lie in $\partial\Omega$ for all $0 < \tau < t < \infty$. Then to solve Eq. (8.0.4) in Ω we only need to prescribe the initial condition (8.0.5).

In later sections we shall encounter conservation laws (8.0.4) with

$$\mathbf{v} = \mathbf{v}(\rho, \mathbf{x}, t), \quad f = f(\rho, \mathbf{x}, t).$$

In such cases one may proceed by successive approximation to solve for ρ_1, $\rho_2, \ldots, \rho_n, \ldots$, where

$$\frac{\partial\rho_{n+1}}{\partial t} + \operatorname{div}\left(\mathbf{v}(\rho_n, \mathbf{x}, t)\rho_{n+1}\right) = f(\rho_n, \mathbf{x}, t).$$

PROBLEM 8.1. Solve, by explicit formula, the conservation law

$$\frac{\partial u}{\partial t} + 2\frac{\partial u}{\partial x} = 0 \quad \text{in} \quad 0 < x < L, \ 0 < t < T,$$

$$u(x,0) = \cos^2(x), \quad 0 < x < L,$$

$$u(0,t) = 1, \quad 0 < t < T$$

for $L = 10$, $T = 5$.

PROBLEM 8.2. Solve, by explicit formula, the conservation law

$$\frac{\partial u}{\partial t} + x\frac{\partial u}{\partial x} = t^2 + 1,$$

$$u(x,0) = \sin(x),$$

$$\text{for} \quad -\infty < x < \infty, \ t > 0.$$

PROBLEM 8.3. Consider the initial value problem (8.0.4), (8.0.5). Prove that if $f(\mathbf{x}, t) \geq 0$, $\rho_0(\mathbf{x}) \geq 0$ for all $\mathbf{x} \in R^n$, $t > 0$, then $\rho(\mathbf{x}, t) \geq 0$ for all $\mathbf{x} \in R^n$, $t > 0$.

8.1. Age Structure of Cells

Eukaryotic cell cycle is divided into four phases, as illustrated schematically in Fig. 8.1 on a scale of 24 h. Synthesis or replication of the DNA occurs in **phase** S. This phase is preceded by **gap phase** G_1 and is followed by **gap phase** G_2. In phase G_1 the cell grows and prepares for entering into the S phase, while in gap phase G_2 the cell, while still growing, checks that the DNA synthesis was done correctly. During the M phase of **mitosis** the nuclear membrane breaks down, duplicate chromosomes migrate to two opposite poles, new nuclear membranes are formed, and two daughter cells split. At check point R_1 the cell decides whether to proceed to the S phase or to go to quiescent state G_0, depending on environmental conditions. The two daughter cells undergo similar life cycle, and so do their descendants. But after a limited number of generations the newborn cells become senescent, they can no longer divide. There are however cells that can divide forever; these are the **stem cells**.

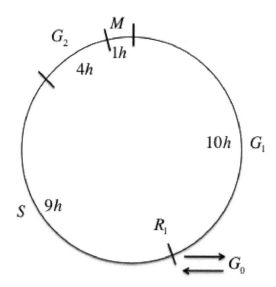

FIGURE 8.1. Schematics of cell cycle

We denote by $p(a, t)$ the population density of cells of age a at time t; here the age of a cell is defined as the time elapsed from the last mitosis. Notice that if we think of a as function of time, then $da/dt = 1$. We denote by $\beta(a)$ the rate of cell division; $\beta(a)$ may vary within the same type of cells. Then, by conservation of cells density,

$$\frac{\partial}{\partial t}p(a, t) + \frac{\partial}{\partial a}p(a, t) = -\beta(a)p(a, t) - \mu p(a, t) \tag{8.1.1}$$

where μ is the death rate of cells. Clearly, $\beta(a) = 0$ if a is too small or too large. When a cell divides into two cells, then the age of the daughter cells is $a = 0$. Hence

$$p(0, t) = 2 \int_0^\infty \beta(a)p(a, t)da.$$

However, a certain fraction q of daughter cells are senescent and will no longer self replicate. Hence the population number density of newly born cells which can self-replicate should actually be given by

$$p(0, t) = 2(1 - q) \int_0^\infty \beta(a)p(a, t)da. \qquad (8.1.2)$$

If we denote by $Q(t)$ the number density of senescent cells at time t, then

$$\frac{d}{dt}Q(t) = 2q \int_0^\infty \beta(a)p(a, t)da - \mu Q(t). \qquad (8.1.3)$$

Here we assumed, for simplicity, that senescent cells' death rate is the same as that of self-replicating cells. If we denote by $P(t) = \int_0^\infty p(a, t)da$ the density of self-replicating cells at time t, then the total density of cells at time t is

$$N(t) = Q(t) + P(t).$$

Cancer cells proliferate faster than normal healthy cells. Some anticancer drugs are aimed at destroying the capability of cells to self-replicate, that is, at increasing the parameter q.

PROBLEM 8.4. Let

$$p(a, 0) = \begin{cases} 1 & \text{for} & 0 < a < 10, \\ 0 & \text{for} & a > 10, \end{cases}$$

$$\beta(a) = \begin{cases} (a - 4)(14 - a) & \text{if} & 4 < a < 14, \\ 0 & \text{if} & a < 4 \quad \text{or} \quad a > 14, \end{cases}$$

and $\mu = 1$. Compute the total population $N(t)$ for $0 < t < 20$ at different efficacy levels of anticancer drug:

$$q = 0.1, \quad q = 0.5, \quad q = 0.8.$$

8.2. Antibiotic Resistance in Hospitals

Antibiotic resistance organisms (AROs) in hospitals pose increasingly serious public health threat. Factors which contribute to the spread of AROs are close living quarters, poor immune system for most patients, and health-care workers who move from one patient to another.

One of the most life-threatening ARO is methicillin-resistant Staphylococcus aureus (MRSA). There are currently several strains of MRSA, with different levels of antibiotic resistance. Here we introduce a simple model with just two bacterial strains, b_1 and b_2, where b_1 is a nonresistant strain and b_2 is drug resistant.

We assume that in each patient there is some bacteria $b = (b_1, b_2)$, and denote by $P(b_1, b_2, t)$, or $P(b, t)$, the population density of patients with bacterial load b at time t. The bacterial load b within each person, and thus also within $P(b, t)$, changes in time with velocity:

$$A_i(b, t) = A_i(b_1, b_2, t) = \frac{db_i(t)}{dt} \quad \text{for strain } i.$$

If the number of patients in the hospital remains constant, then, by conservation law,

$$\frac{\partial P(b, t)}{\partial t} + \sum_{i=1}^{2} \frac{\partial}{\partial b_i} [A_i(b, t) P(b, t)] = 0. \tag{8.2.1}$$

Next we introduce a simple model for $A_i(b, t)$, ignoring the effect of healthcare workers:

$$A_1 = \frac{db_1}{dt} = \lambda_1 b_1 - \frac{\nu_1 b_1^2}{1 + b_1^2} - \sigma_1(t) b_1 - \frac{M b_1}{1 + \sigma_1(t)},$$

$$A_2 = \frac{db_2}{dt} = \lambda_2 b_2 - \frac{\nu_2 b_2^2}{1 + b_2^2} - \sigma_2(t) b_2 + \frac{M b_2}{1 + \sigma_1(t)}.$$

Here λ_i is the rate of bacterial growth b_i, $\nu_i b_i^2 / (1 + b_i^2)$ is the patient immune response which kills bacteria by an enzymatic process modeled by Hill dynamics, $\sigma_i(t)$ is the killing rate of b_i by antibiotic drug, and $M b_2 / (1 + \sigma_1(t))$ in A_2 represents the mutation from b_1 to b_2. Notice that the mutation rate increases if the dose of antibiotic treatment decreases. The total amount of bacterial strain b_i in the hospital at time t is given by the formula

$$Q_i(t) = \int_{R_2^+} b_i P(b, t) db$$

where

$$R_2^+ = \{(b_1, b_2), 0 < b_1 < \infty, 0 < b_2 < \infty\}.$$

We take $\sigma_i(t) = \alpha_i h(t)$ where $h(t)$ is the schedule by which drug is administered, that is,

$$h(t) = \begin{cases} 1 & \text{if drug is given at time } t, \\ 0 & \text{if no drug is given at time } t. \end{cases}$$

Since b_2 is antibiotic resistant, $\alpha_2 < \alpha_1$ and $\nu_2 < \nu_1$. Suppose a fixed amount of drug is given over a period T, that is,

$$\int_0^T h(t) dt = K.$$

The question for doctors is what is the best choice for $h(t)$ so that $Q_2(T)$ is minimum.

PROBLEM 8.5. Take $\lambda_1 = 2.7$, $\lambda_2 = 0.9$, $\nu_1 = 3$, $\nu_2 = 1$, $\alpha_1 = 2$, $\alpha_2 = 0.2$, $T = 28$, and

$$P(b_1, b_2, 0) = \begin{cases} 3 & \text{if} \quad 1 < b_1 + b_2 < 2, \\ 0 & \text{if} \quad b_1 + b_2 < 1 \text{ or } b_1 + b_2 > 2. \end{cases}$$

Consider two strategies:

    (i) $h(t) = 5$ if $0 < t < 14$, $h(t) = 0$ if $14 < t < 28$;
    (ii) $h(t) = 10$ if $0 < t < 7$, $h(t) = 0$ if $7 < t < 28$.

Compute $Q_2(T)$ for both strategies, and determine which of the two treatments yields a smaller $Q_2(T)$.

8.3. Numerical Methods for Hyperbolic Equations

Developing numerical methods for hyperbolic conservation laws [13, 16] which can capture the unique physically relevant solutions (the so-called "viscosity solutions") is an important area of numerical analysis. Here we introduce two simple first-order methods, Upwind scheme and Lax–Friedrichs scheme, which can be written in the discrete conservation form.

The general nonlinear conservation equation is

$$u_t + (f(u))_x = 0. \tag{8.3.1}$$

Many schemes are built based on a cell average idea. We discretize the $x - t$ plane by choosing a mesh width $h = \Delta x$ and a time step $k = \Delta t$, and define the discrete mesh points (x_j, t_n) by

$$\begin{aligned} x_j &= jh, & j &= \dots, -1, 0, 1, 2, \dots \\ t_n &= nk, & n &= 0, 1, 2, \dots \end{aligned}$$

Integrating (8.3.1) over $x_{j-\frac{1}{2}} \leq x \leq x_{j+\frac{1}{2}}$, $t_n \leq t \leq t_{n+1}$ and dividing h give

$$\frac{1}{h} \int_{x_{j-\frac{1}{2}}}^{x_{j+\frac{1}{2}}} u(x, t_{n+1}) dx = \frac{1}{h} \int_{x_{j-\frac{1}{2}}}^{x_{j+\frac{1}{2}}} u(x, t_n) dx$$

$$- \frac{k}{h} \left(\frac{1}{k} \int_{t_n}^{t_{n+1}} f\left(u(x_{j+\frac{1}{2}}, t)\right) dt - \frac{1}{k} \int_{t_n}^{t_{n+1}} f\left(u(x_{j-\frac{1}{2}}, t)\right) dt \right). \tag{8.3.2}$$

Thus

$$\bar{u}_{I_j}^{n+1} = \bar{u}_{I_j}^{n} - \frac{k}{h} \left(\frac{1}{k} \int_{t_n}^{t_{n+1}} f\left(u(x_{j+\frac{1}{2}}, t)\right) dt - \frac{1}{k} \int_{t_n}^{t_{n+1}} f\left(u(x_{j-\frac{1}{2}}, t)\right) dt \right)$$

where $\bar{u}_{I_j}^{n}$ indicates the cell average in $I_j = [x_{j-\frac{1}{2}}, x_{j+\frac{1}{2}}]$ at time $t = nk$. Numerically, the discrete conservation form

$$U_j^{n+1} = U_j^n - \frac{k}{h} \left(F(U_{j-p}, \dots, U_{j+q}) - F(U_{j-p-1}, \dots, U_{j+q-1}) \right) \tag{8.3.3}$$

for some function F of $p + q + 1$ arguments is used to solve Eq. (8.3.1). Comparing Eq. (8.3.2) and (8.3.3), we know that the function $F(U_{j-p}, \dots, U_{j+q})$

should give good approximation of average flux at $x_{j+\frac{1}{2}}$ in the time interval $[t_n, t_{n+1}]$,

$$\frac{1}{k} \int_{t_n}^{t_{n+1}} f\left(u(x_{j+\frac{1}{2}}, t)\right) dt,$$

in order to provide reasonable numerical solutions. Thus the function F is called "*numerical flux*." The choice of p and q depends on how accurate the schemes one wants to design. The simple three-stencil schemes are schemes with $p = 0$ and $q = 1$, which means that $F(U_j, U_{j+1})$ is used to approximate

$$\frac{1}{k} \int_{t_n}^{t_{n+1}} f\left(u(x_{j+\frac{1}{2}}, t)\right) dt,$$

while $F(U_{j-1}, U_j)$ is used to approximate

$$\frac{1}{k} \int_{t_n}^{t_{n+1}} f\left(u(x_{j-\frac{1}{2}}, t)\right) dt,$$

and then

$$U_j^{n+1} = U_j^n - \frac{k}{h}\left(F(U_j^n, U_{j+1}^n) - F(U_{j-1}^n, U_j^n)\right).$$

The Upwind scheme suggests to use the numerical flux:

$$F_{Upwind}(U_j^n, U_{j+1}^n) = \begin{cases} f(U_j^n) & \text{if } \frac{f(U_{j+1}^n) - f(U_j^n)}{U_{j+1} - U_j} > 0, \\ f(U_{j+1}^n) & \text{if } \frac{f(U_{j+1}^n) - f(U_j^n)}{U_{j+1}^n - U_j^n} \leq 0. \end{cases}$$

This scheme numerically mimics the theoretical characteristic direction to propagate the information. The Lax–Friedrichs scheme suggests to use the numerical flux

$$F_{LF}(U_j^n, U_{j+1}^n) = \frac{1}{2}\left(f(U_j^n) + f(U_{j+1}^n) - \alpha(U_{j+1}^n - U_j^n)\right),$$

where

$$\alpha = \max_u |f'(u)|$$

is introduced to ensure stability. Thus

$$U_j^{n+1} = U_j^n - \frac{k}{h}\left(\frac{1}{2}\left(f(U_j^n) + f(U_{j+1}^n) - \alpha(U_{j+1}^n - U_j^n)\right)\right.$$
$$\left. - \frac{1}{2}\left(f(U_{j-1}^n) + f(U_j^n) - \alpha(U_j^n - U_{j-1}^n)\right)\right),$$

which gives the update formula

$$U_j^{n+1} = U_j^n - \frac{1}{2}\frac{k}{h}\left(f(U_{j+1}^n) - f(U_{j-1}^n) - \alpha(U_{j+1}^n - 2U_j^n + U_{j-1}^n)\right).$$

In Algorithm 9–13, the MATLAB codes of Upwind and Lax–Friedrichs schemes are provided for Problem 8.1. The numerical solutions for Upwind and Lax–Friedrichs schemes are shown in Figs. 8.2 and 8.3, respectively. Both of the schemes are of first-order accurate. The above schemes are called *three-stencil schemes* because U_{j-1}^n, U_j^n, and U_{j+1}^n are involved.

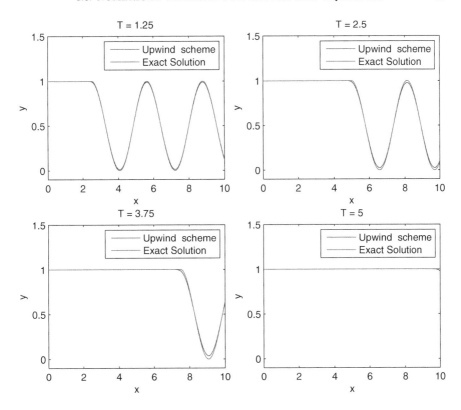

FIGURE 8.2. The solution of Upwind scheme for problem 8.1

PROBLEM 8.6. Solve the system

$$\frac{\partial u}{\partial t} + 2\frac{\partial u}{\partial x} = -u + v, \quad \frac{\partial v}{\partial t} - \frac{\partial v}{\partial x} = u - v \quad \text{in} \quad 0 < x < L, \ 0 < t < T,$$

$$u(x,0) = \cos^2(x), v(x,0) = 2\sin^2(x), \quad 0 < x < L,$$

$$u(0,t) = 1, v(L,t) = 2, \quad 0 < t < T$$

for $L = \frac{3\pi}{2}$, $T = 10$.

PROBLEM 8.7. Let $v(x,t) = x^2 + t$, $f(x) = x^2$ and consider the (hyperbolic) conservation law

$$\frac{\partial u}{\partial t} + v\frac{\partial u}{\partial x} = f \quad \text{in} \quad 0 < x < s(t),$$

with

$$u(0,t) = 1, \quad t > 0,$$
$$u(x,0) = 1 + x^2, \quad 0 < x < s(t)$$

where

$$s(0) = 2, \quad \dot{s}(t) = v(s(t),t).$$

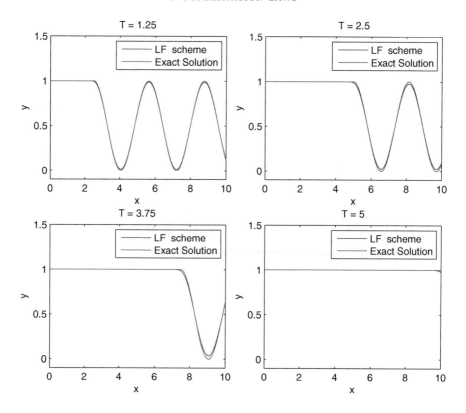

FIGURE 8.3. The solution of Lax–Friedrichs scheme for problem 8.1

Solve u by the method of characteristics. Note that since the boundary $x = s(t)$ is a characteristic curve, no boundary conditions are needed on $x = s(t)$.

PROBLEM 8.8. Solve the equation

$$\frac{\partial u}{\partial t} + \frac{\partial}{\partial x}(vu) = f$$

with the same data as in Problem 8.7.

Algorithm 9 main_Upwind_LF.m

```
% main_Upwind_LF.m
% Upwind and Lax-Friedrichs schemes for the transport equation
"u_t+(f(u))_x==0".
% with flux function 2*u and
% the initial condition:
% cos^2(x) for o<x<L, u(0,t) = 1, 0<t<T
clear;
% Set up domain in space and time
L = 10; T = 5;
% Set up parameters for the solver.
vmax = 2; lambda = .5/vmax;
N = 1000; % number of intervals in x discretization
h = L/N;
X = 0:h:L;
k = lambda*h; % time step
nr_iter = ceil(T/k); %number of steps in time
k = T/nr_iter; % adjust time step
plot_iter = floor(1.25/k);
scheme = 'LF' % LF or Upwind
% Set up the initial iteration.
U = IC1(X);
figure(1);winsize = get(gca,'Position');set(gca,'Position',winsize*0.4)
for j=1:nr_iter
ct = j*k; % current time
% compute exact solution
Exactsol = exact1(X,ct);
switch scheme
case 'LF'
% LF Scheme
U(1) = bd1(ct); % boundary condition
FpLF = 0.5*(flux1(U(2:N))+flux1(U(3:N+1))-vmax*(U(3:N+1)-U(2:N)));
FmLF = 0.5*(flux1(U(1:N-1))+flux1(U(2:N))-vmax*(U(2:N)-U(1:N-1)));
U(2:N) = U(2:N)-lambda*(FpLF-FmLF);
U(N+1) = U(N+1)-lambda*(flux1(U(N+1))-flux1(U(N)));
case 'Upwind' % Upwind Scheme
U(1) = bd1(ct); % boundary condition
FpUp = flux1(U(2:N+1));
FmUp = flux1(U(1:N));
U(2:N+1) = U(2:N+1)-lambda*(FpUp-FmUp);
otherwise
error('no scheme defined')
end
```

(continued)

Algorithm 9 (continued)

```
if mod(j,plot_iter)==0
plot(X,U,'b',X,Exactsol,'r');
legend([scheme ' scheme'],'Exact Solution')
title(['T = ' num2str(ct)]);
xlabel('x'); ylabel('y')
axis([0 L -0.1 1.5])
print(gcf,'-depsc',[scheme 'scheme_T_' num2str(ct) '.eps' ])
drawnow;
end
end
```

Algorithm 10 exact1.m

```
function val = exact1(x,t);
val = ((x-2*t)>0).*IC1(x-2*t)+1*((x-2*t)<=0);
```

Algorithm 11 flux1.m

```
function val = flux1(u) % this is flux function
val = 2*u;
```

Algorithm 12 IC1.m

```
function val = IC1(x);
val = cos(x).^2;
```

Algorithm 13 bd1.m

```
function val = bd1(t);
val = 1;
```

CHAPTER 9

Neurofilaments Transport in Axon

Most axonal proteins are synthesized in the nerve cell body and are then transported along the axon. One type of cargo structures is neurofilaments, neuron-specific cytoskeleton polymers that function as space-filling structure in axon. Each such cargo is attached to a motor protein that travels along microtubules. This piggybacking transport is illustrated in Fig. 9.1.

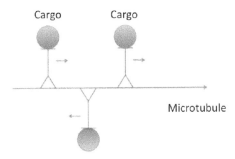

FIGURE 9.1. Cargo-carrying motor proteins travel on microtubules

Studies of populations of neurofilaments in vivo on timescale of weeks and months using radioscopic pulse-labeling have demonstrated that these cytoskeletal polymers move along axons in slow average rates of about 0.002–0.035 μm/s, i.e., about 0.25–3 mm/day. More recently it became technically possible to observe movement of neurofilaments directly in cultured nerve cells on a timescale of seconds to minutes, using fluorescence microscopy. These observations have demonstrated that the movement of neurofilaments is not that slow; neurofilaments actually move at fast rates (about 0.5 μm/s), and the movements are also infrequent, bidirectional, and highly asynchronous.

The question then arose how to explain the movement of neurofilaments so that the in vivo measurements are in agreement with the in vitro measurements. This is an important question since defects in the transport of neurofilaments in neurons may contribute to their accumulation in some neurodegenerative diseases, such as ALS.

© Springer International Publishing Switzerland 2014
A. Friedman, C.-Y. Kao, *Mathematical Modeling of Biological Processes*, Lecture Notes on Mathematical Modelling in the Life Sciences, DOI 10.1007/978-3-319-08314-8_9

In the present chapter we develop a dynamical system model, based on a mass conservation law, for the transport of neurofilaments in axons in vivo, based on the fast timescale observations in cultured nerve cells. We then show by simulations, that the model can match the population profiles of pulse-labeled neurofilaments obtained experimentally in rat and mouse lumbar spinal motor neurons. The model was developed in [6], the measurements of neurofilaments in cultured nerve cells were reported in [21] and the references therein, and the experiments with rat and mouse were reported in [22].

To begin the model we first need to recall some facts regarding the mec - hanics of the movement of the motor proteins which carry the cargo of neurofilaments along microtubules. The microtubules are formed by proteins and are linearly assembled along the axon. The cargo gets attached on one end of a motor protein, while the other end of the motor protein is loosely attached to the microtubule.

Kinesin motor proteins move forward toward the synaptic terminal (anterograde motion), and **dynein** motor proteins move backward (retrograde motion). We assume that neurofilaments are only capable of movement in the longitudinal dimension of the axon when they are on track along the microtubule, but they can switch on and off track.

When off track, neurofilaments pause for long periods until they get back on track. When on track, neurofilaments alternate between short bouts of movement and short pauses. Thus we divide the on track population of neurofilaments into those that are moving and those that are making a short pause. We designate separately the population of anterograde moving neurofilaments and the population of retrograde moving neurofilaments. In this manner, we divide the neurofilaments into five populations:

u_2 : neurofilaments bound to anterograde motors, moving anterogradely, on track,

u_1 : neurofilaments bound to anterograde motors, pausing, on track,

u_0 : neurofilaments bound to anterograde or retrograde, motors, pausing, off track,

u_{-1} : neurofilaments bound to retrograde motors, pausing, on track,

u_{-2} : neurofilaments bound to retrograde motors, moving retrogradely, on track.

For simplicity we shall also denote by u_{-2}, u_{-1}, u_0, u_1, u_2 the concentrations of these five populations of neurofilaments along the axon.

We assume that neurofilaments can reverse direction (i.e., switch motors) when they are pausing off track. Thus we obtain the following diagram of possible transitions between the five neurofilament populations:

$$u_{-2} \leftrightarrows u_{-1} \leftrightarrows u_0 \rightleftarrows u_1 \rightleftarrows u_2$$

If we denote by $k_{i,j}$ the rate of change from u_i to u_j, then we have

$$
u_{-2} \underset{k_{-2,-1}}{\overset{k_{-1,-2}}{\leftrightarrows}} u_{-1} \underset{k_{-1,0}}{\overset{k_{0,-1}}{\leftrightarrows}} u_0 \underset{k_{1,0}}{\overset{k_{0,1}}{\rightleftarrows}} u_1 \underset{k_{2,1}}{\overset{k_{1,2}}{\rightleftarrows}} u_2.
$$

These relations, together with the mass conservation law, yield the following dynamical system for $0 < x < L$, where L is the length of the axon:

$$
\begin{aligned}
\frac{\partial u_2}{\partial t} &= -v_A \frac{\partial u_2}{\partial x} + k_{1,2}u_1 - k_{2,1}u_2, \\
\frac{\partial u_1}{\partial t} &= k_{2,1}u_2 + k_{0,1}u_0 - (k_{1,2} + k_{1,0})u_1, \\
\frac{\partial u_0}{\partial t} &= k_{1,0}u_1 + k_{-1,0}u_{-1} - (k_{0,1} + k_{0,-1})u_0, \\
\frac{\partial u_{-1}}{\partial t} &= k_{0,-1}u_0 + k_{-2,-1}u_{-2} - (k_{-1,0} + k_{-1,-2})u_{-1}, \\
\frac{\partial u_{-2}}{\partial t} &= v_R \frac{\partial u_{-2}}{\partial x} + k_{-1,-2}u_{-1} - k_{-2,-1}u_{-2},
\end{aligned}
\tag{9.0.1}
$$

where, in vivo, v_A is the velocity of neurofilaments moving anterograde and v_R is the velocity of neurofilaments moving retrograde. It is known experimentally [21] that

$$
v_A = 0.56\mu\text{m/s}, \quad v_R = 0.62\mu\text{m/s}.
$$

The system (9.0.1) models the movement of neurofilaments in vivo, but in order to use it to compute the densities u_i we need to find the parameters. The search for parameters is a typical difficulty one encounters in making mathematical models useful. So far we only noted that v_A and v_R have been determined experimentally, but what about the $k_{i,j}$?

We shall now show how to interpret the measurements done in vitro in [21] in order to determine most of the $k_{i,j}$. Since these measurements are done in a short period of few seconds, they do not change the total populations u_i while the measurements are being conducted. Suppose we observe that, while n neurofilaments move retrograde (from u_{-2} to u_{-1}), $n\alpha$ neurofilaments move retrograde (from u_{-1} to u_{-2}). Since in vitro the exchanges between u_{-1} and u_{-2} take place with the same velocity, this means that there are α times more neurofilaments in u_{-1} than in u_{-2}. Hence we deduce that in vitro

$$
\frac{u_{-1}}{u_{-2}} = \alpha = \frac{k_{-2,-1}}{k_{-1,-2}}.
\tag{9.0.2}
$$

Another observation is that the movement of neurofilaments seen in vitro during the few seconds of counting them cross the screen is "fast," in the sense that each $k_{i,j}$ is a large number. Another way of looking at it is that while in vivo the movement is considered only over a period of weeks or months, in the measurements done in vitro, with the in vitro populations u_i, the exchanges over the period of a few seconds, among the various populations, are "fast." Hence we assume that, $k_{-1,-2}$ is a large number, λ, say $10 < \lambda < 100$.

Recalling (9.0.2) we conclude that in vitro

$$\frac{u_{-1}}{u_{-2}} = \frac{\alpha\lambda}{\lambda} \quad \text{and} \quad k_{-1,-2} = \lambda, \quad k_{-2,-1} = \alpha\lambda. \tag{9.0.3}$$

Similarly we have

$$\frac{u_0}{u_{-1}} = \frac{\beta\lambda}{\lambda} \quad \text{and} \quad k_{0,-1} = \lambda, \quad k_{-1,0} = \beta\lambda, \tag{9.0.4}$$

where, for simplicity, we take the same λ, and

$$\frac{u_0}{u_1} = \frac{\gamma\lambda}{\lambda} \quad \text{and} \quad k_{0,1} = \lambda, \quad , k_{1,0} = \gamma\lambda, \tag{9.0.5}$$

and

$$\frac{u_1}{u_2} = \frac{\delta\lambda}{\lambda} \quad \text{and} \quad k_{1,2} = \lambda, \quad k_{2,1} = \delta\lambda. \tag{9.0.6}$$

If we look at the data in [21], we notice that, in most cases, individual neurofilaments moving anterogradely switch back and forth between populations u_1 (short pause) and u_2 (moving). If individual neurofilaments moving anterograde make more short pauses, then u_2 decreases and u_1 increases. The ratio u_1/u_2 depends on the time spent pausing and the time spent in processive motion. From the experimental measurements in [21] we deduce that

$$\frac{u_1}{u_2} = \frac{u_{-1}}{u_{-2}} = \frac{67}{33}.$$

Hence

$$\alpha = \delta = \frac{67}{33}. \tag{9.0.7}$$

Also the ratio between neurofilaments moving anterograde and neurofilaments moving retrograde is 69/31, i.e.,

$$\frac{u_1 + u_2}{u_{-1} + u_{-2}} = \frac{69}{31}.$$

Using (9.0.7) and (9.0.2)–(9.0.6) we find that

$$\beta = \frac{69}{31}\gamma. \tag{9.0.8}$$

Thus β is determined if γ is known. So now we have determined all the parameters $k_{i,j}$ provided we prescribe γ.

PROBLEM 9.1. Consider the following movement of two kinds of cargo: u moving anterograde and v moving retrograde:

$$\frac{\partial u}{\partial t} + \frac{\partial u}{\partial x} = v - u,$$
$$\frac{\partial v}{\partial t} - \frac{\partial v}{\partial x} = u - v,$$

for $-\infty < x < \infty, 0 < t < \infty$, such that

$$u(x,0) \geq 0, v(x,0) \geq 0 \quad \text{for} \ -\infty < x < \infty.$$

Prove that

$$u(x,t) \geq \int_0^t ds \int_0^s e^{-(t-\sigma)} u(x-t+2s-\sigma,\sigma) d\sigma.$$

[Hint: Write the equations for u and v in the form

$$\frac{\partial}{\partial t}\left(e^t u\right) + \frac{\partial}{\partial x}\left(e^t u\right) = e^t v, \quad \frac{\partial}{\partial t}\left(e^t v\right) - \frac{\partial}{\partial x}\left(e^t v\right) = e^t u.$$

The characteristic curve of the first equation with $\xi(t) = x$ is $\xi(\tau) = \tau + x - t$, and

$$\frac{d}{d\tau}\left[e^\tau u(x-t+\tau,\tau)\right] = e^\tau v(x-t+\tau,\tau).$$

Integrate over $0 \leq \tau \leq t$ to get

$$u(x,t) \geq \int_0^t e^{-(t-s)} v(x-t+s,s) ds.$$

Now repeat this procedure with the second equation to estimate $v(x-t+s,s)$, using the characteristic curve $\xi(\tau) = -\tau + 2s + x - t$:

$$v(x-t+s,s) \geq \int_0^s e^{-(s-\sigma)} u(x-t+2s-\sigma,\sigma) d\sigma,$$

and then substitute into the first inequality.]

9.1. Numerical Computation of Neurofilaments Transport in Axon

To solve the system of Eqs. (9.0.1) numerically, we use the Upwind method introduced in Sect. 8.3 for the advection term and use the current time step value for its own population and previous time step for other populations for the reaction term. We discretize the $x - t$ plane by choosing a mesh width Δx and a time step Δt and define the discrete mesh points (x_j, t_n) by

$$\begin{aligned}
x_j &= j\Delta x, & j &= \ldots, -1, 0, 1, 2, \ldots \\
t_n &= n\Delta t, & n &= 0, 1, 2, \ldots
\end{aligned}$$

This leads to the discretization in the following form:

$$\begin{aligned}
\frac{u_{2,j}^{n+1} - u_{2,j}^n}{\Delta t} &= -v_A \frac{u_{2,j}^n - u_{2,j-1}^n}{\Delta x} + k_{1,2} u_{1,j}^n - k_{2,1} u_{2,j}^{n+1}, \\
\frac{u_{1,j}^{n+1} - u_{1,j}^n}{\Delta t} &= k_{2,1} u_{2,j}^n + k_{0,1} u_{0,j}^n - (k_{1,2} + k_{1,0}) u_{1,j}^{n+1}, \\
\frac{u_{0,j}^{n+1} - u_{0,j}^n}{\Delta t} &= k_{1,0} u_{1,j}^n + k_{-1,0} u_{-1,j}^n - (k_{0,1} + k_{0,-1}) u_{0,j}^{n+1}, \qquad (9.1.1) \\
\frac{u_{-1,j}^{n+1} - u_{-1,j}^n}{\Delta t} &= k_{0,-1} u_{0,j}^n + k_{-2,-1} u_{-2,j}^n - (k_{-1,0} + k_{-1,-2}) u_{-1,j}^{n+1}, \\
\frac{u_{-2,j}^{n+1} - u_{-2,j}^n}{\Delta t} &= v_R \frac{u_{-2,j+1}^n - u_{-2,j}^n}{\Delta x} + k_{-1,-2} u_{-1,j}^n - k_{-2,-1} u_{-2,j}^{n+1}.
\end{aligned}$$

The advantage of this choice is that the population will remain nonnegative as long as the initial population is nonnegative when the CFL constrain is satisfied, e.g.,

$$\Delta t \le \frac{0.5\Delta x}{\min\{v_A, v_R\}}.$$

Furthermore, the discretization leads to a simple explicit update formula:

$$
\begin{aligned}
u_{2,j}^{n+1} &= \left\{u_{2,j}^n + \Delta t \left(-v_A \frac{u_{2,j}^n - u_{2,j-1}^n}{\Delta x} + k_{1,2} u_{1,j}^n\right)\right\} / (1 + \Delta t k_{2,1}),\\
u_{1,j}^{n+1} &= \left\{u_{1,j}^n + \Delta t \left(k_{2,1} u_{2,j}^n + k_{0,1} u_{0,j}^n\right)\right\} / (1 + \Delta t (k_{1,2} + k_{1,0})),\\
u_{0,j}^{n+1} &= \left\{u_{0,j}^n + \Delta t \left(k_{1,0} u_{1,j}^n + k_{-1,0} u_{-1,j}^n\right)\right\} / (1 + \Delta t (k_{0,1} + k_{0,-1})),\qquad (9.1.2)\\
u_{-1,j}^{n+1} &= \left\{u_{-1,j}^n + \Delta t \left(k_{0,-1} u_{0,j}^n + k_{-2,-1} u_{-2,j}^n\right)\right\} / (1 + \Delta t (k_{-1,0} + k_{-1,-2})),\\
u_{-2,j}^{n+1} &= \left\{u_{-2,j}^n + \Delta t \left(v_R \frac{u_{-2,j+1}^n - u_{-2,j}^n}{\Delta x} + k_{-1,-2} u_{-1,j}^n\right)\right\} / (1 + \Delta t k_{-2,-1}).
\end{aligned}
$$

In [6], the simulation of the model was done with initial conditions

$$u_0(x,0) = A e^{-(x-x_0)^2/M}, \quad u_j(x,0) = 0 \text{ for all other } j.$$

where A, M, and x_0 are chosen which fit best to the experimental measurements in week 1 in [22] and $\lambda = 100$, and then choose γ such that the profile of the total population of neurofilaments best fits the experimental results of [22] for week 3. The parameter $\gamma = 19.9$ was found to give a good fit as shown in Fig. 9.2. Similar simulations have been carried out in [6] to fit experimental results in other weeks and for other neurons, but each week and each neuron require a different choice of γ; variation in γ may be due to the fact that transport kinetics do vary spatially and/or temporally in some axons.

In Algorithm (14)–(15), we demonstrate the simulation of the discretization (9.1.2) with $A = 0.4$, $x0 = 5$, $M = 15$ and show the numerical results in Fig. 9.3. For ease of visualization, the solutions $(25u_2, 5u_1, u_0, 5u_{-1}, 25u_{-2})$ are shown so that the populations are in the comparable size.

PROBLEM 9.2. Take A, M, x_0, and L that fit to the blue profile (week 1) in Fig. 9.2 and simulate the model (9.0.1) with the same parameters. Choose $\lambda = 20$ and compare the profile in week 3 to the one in Fig. 9.2.

PROBLEM 9.3. Consider the system (9.0.1) for $0 < x < L$ with the parameters specified above, and

$$u_0(x,0) = e^{-x^2}, u_j(x,0) = 0 \quad \text{if} \quad j \ne 0 \quad (0 < x < L),$$

$$u_2(0,t) = 1/e, \quad u_{-2}(L,t) = 0, \quad L = 60mm, \quad \text{and} \quad \gamma = 20.$$

Compute the total population $u(x,t) = \sum_{j=-2}^{j=2} u_j(x,t)$ and verify (by taking $t = 2, 3, 4, 5, 6$ weeks) that it has the profile of an approximate traveling wave: the wave travels at nearly constant velocity, but the front spreads.

PROBLEM 9.4. Repeat the calculations with $\gamma = 10$ and $\gamma = 50$.

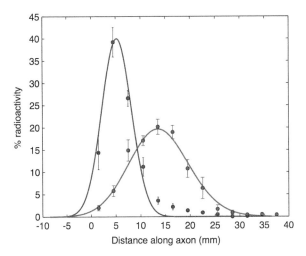

FIGURE 9.2. Population profiles of neurofilament protein in mouse lumbar ventral root and sciatic nerve: the *dots* represent the experimental data; *blue* is week 1, and *red* is week 3. Each point is the mean of $3 - 5$ nerves. The *bars* represent standard error of the mean. The *blue* curve is the initial condition for the model; the *red curve* then follows from the dynamical system model

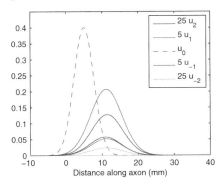

FIGURE 9.3. For ease of visualization, the solutions ($25u_2$, $5u_1, u_0, 5u_{-1}, 25u_{-2}$) are shown so that the populations are in the comparable size. The initial population of u_0 is indicated by the *red dashline*

Algorithm 14 main_neurofilaments.m

```
% main_Neurofilaments.m
% Use Upwind scheme to solve the model for neurofilaments transport in
Axon clear;
% Set up parameters for the solver.
vA = 0.56*0.001*60*60*24; % change the speed to mm/day
vR = 0.62*0.001*60*60*24;
CF = .5/max(vA,vR); % Use CFL condition to determine the time step
A = 0.4; x0 = 5; M = 15;
lambda = 100; gamma = 19.9;
beta = 69/31*gamma;
alpha = 67/33;
delta = alpha;
km2m1 = alpha*lambda; km1m2 = lambda;
km10 = beta*lambda; k0m1 = lambda;
k10 = gamma*lambda; k01 = lambda;
k21 = delta*lambda; k12 = lambda;
% Set up domain in space and time
T = 2*7; % 14 days
N = 400; % number of intervals in x discretization
xmin = -10; xmax = 40; h = (xmax-xmin)/N; X = xmin:h:xmax;
k = CF*h;
% time step
nr_iter = T*ceil(1/k); %number of steps in time k = T/nr_iter;
% adjust time step
plot_iter = floor(0.5/k);
% Set up the initial iteration.
U = IC2(X,A,x0,M);
figure(1); plot(X,25*U(1,:),X,5*U(2,:),X,U(3,:),'r–',X,5*U(4,:),X,25*U(5,:))
legend('25 u_2','5 u_1','u_0','5 u_{-1}','25 u_{-2}')
xlabel('Distance along axon (mm)')
for j=1:nr_iter
ct = j*k;
% current time
% Upwind Scheme
U(1,2:end) =
(U(1,2:end)+k*(-vA*(U(1,2:end)-
U(1,1:end-1))/h+k12*U(2,2:end)))/(1+k21*k);
U(2,1:end) = (U(2,1:end)+
k*(k21*U(1,1:end)+k01*U(3,1:end)))/(1+(k12+k10)*k);
U(3,1:end) =
(U(3,1:end)+k*(k10*U(2,1:end)+km10*U(4,1:end)))/(1+(k01+k0m1)*k);
```

(continued)

Algorithm 14 (continued)

```
U(4,1:end) = (U(4,1:end)+k*(k0m1*U(3,1:end)+km2m1*U(5,1:end)))/
            (1+(km10+km1m2)*k);
U(5,1:end-1) = (U(5,1:end-1)+k*(vR*(U(5,2:end)-U(5,1:end-
1))/h+km1m2*U(4,1:end-1)))/(1+km2m1*k);
% boundary conditions
U(1,1) = 0; U(5,end) = 0;
if mod(j,plot_iter)==0
figure(2); plot(X,25*U(1,:),X,5*U(2,:),X,U(3,:),X,5*U(4,:),X,25*U(5,:))
legend('25 u_2','5 u_1','u_0','5 u_{-1}','25 u_{-2}')
title(['T = ' num2str(ct)]);xlabel('x'); ylabel('y') drawnow;
end
end
figure(1);hold on;
plot(X,25*U(1,:),X,5*U(2,:),X,U(3,:),X,5*U(4,:),X,25*U(5,:))
axis([-10 40 0 0.45])
```

Algorithm 15 IC2.m

```
function val = IC2(x,A,x0,M);
val = [zeros(size(x));zeros(size(x));
A*exp(-(x-x0).^2/M);zeros(size(x));zeros(size(x))];
```

CHAPTER 10

Diffusion and Chemotaxis

Diffusion is the movement of particles from high concentration to low concentration. It arises from the fact that all particles are constantly moving in random directions. Biological organisms, for instance, cells, move not only at random but also in response to the environment. This response often involves directed movement toward external stimulus or away from it. Such a movement is call "taxis." **Chemotaxis** is a movement in response to chemical gradient, and **haptotaxis** is a response to adhesive gradient. Biological examples of chemotaxis and haptotaxis abound. In this chapter we give two examples of chemotaxis. The first example has to do with movement of amoebas during one phase of their lifecycle. The second example is concerned with the movement of endothelial cells in response to stimulus secreted by tumor cells. But first we need to develop general mathematical models of diffusion and chemotaxis.

Consider a particle which moves along a straight line by making steps of size Δx at discrete time intervals Δt, and set $x_i = i \Delta x$ $(i = 0, \pm 1, \pm 2, \ldots)$, $t_j = j \Delta t$ $(j = 0, 1, 2, \ldots)$. We assume that the particle is initially at $x = 0$, and we denote by $u_i(t_j)$ the probability that the particle will be at x_i in time t_j. We denote by T_i^+ (T_i^-) the probability of jumping from x_i to x_{i+1} (x_{i-1}). Then, for any $t = t_j$,

$$u_i(t + \Delta t) - u_i(t) = T_{i-1}^+ u_{i-1}(t) + T_{i+1}^- u_{i+1}(t) - (T_i^+ + T_i^-) u_i(t) \quad (10.0.1)$$

The system (10.0.1) will enable us, by going to the limit, with $\Delta t \to 0$, $\Delta x \to 0$, to obtain several different processes that occur in biology. The limit equations will be derived for a function $u(x, t)$ with $u(x_i, t_j) = u_i(t_j)$.

We begin with the case $T_i^{\pm} = \alpha$. Then, for any $t = t_j$, $x = x_i$, we obtain from (10.0.1)

$$u(x, t + \Delta t) - u(x, t) = \alpha \left[u(x - \Delta x, t) + u(x + \Delta x, t) - 2u(x, t) \right]$$

$$= \alpha (\Delta x)^2 \frac{\partial^2 u(x, t)}{\partial x^2} + O\left((\Delta x)^3\right),$$

and the left-hand side is equal to

$$\frac{\partial u(x, t)}{\partial t} \Delta t + O\left((\Delta t)^2\right).$$

© Springer International Publishing Switzerland 2014
A. Friedman, C.-Y. Kao, *Mathematical Modeling of Biological
Processes*, Lecture Notes on Mathematical Modelling in the Life
Sciences, DOI 10.1007/978-3-319-08314-8_10

If we assume that

$$\lim_{\Delta t \to 0} \alpha \frac{(\Delta x)^2}{\Delta t} \to D, \quad D > 0,$$

then we obtain the diffusion equation

$$\frac{\partial u}{\partial t} = D \frac{\partial^2 u}{\partial x^2}. \tag{10.0.2}$$

Note that the probability of finding the particle in an interval $a < x < b$ is

$$\int_a^b u(x, t) dx. \tag{10.0.3}$$

If all particles in a chemical species make the same random walk, then the concentration $u(x,t)$ of the chemical species will satisfy the same equation (10.0.2), and the total mass in any interval $a < x < b$ is then given by (10.0.3).

Consider next the case

$$T_i^+ = \alpha, \quad T_i^- = \beta, \quad \alpha \neq \beta.$$

Then the right-hand side of (10.0.1) at $t = t_j$ becomes

$$\alpha u(x - \Delta x, t) + \beta u(x + \Delta x, t) - (\alpha + \beta) u(x, t),$$

which, by Taylor's formula, is equal to

$$-(\alpha - \beta)(\Delta x) \frac{\partial u}{\partial x}(x, t) + \frac{1}{2}(\alpha + \beta)(\Delta x)^2 \frac{\partial^2 u(x, t)}{\partial x^2} + O\left((\Delta x)^3\right).$$

Assuming that

$$\lim_{\Delta t \to 0} (\alpha - \beta) \frac{\Delta x}{\Delta t} = v,$$

$$\lim_{\Delta t \to 0} (\alpha + \beta) \frac{(\Delta x)^2}{\Delta t} = 2D, \quad D > 0,$$

we get the diffusion–advection equation

$$\frac{\partial u}{\partial t} + v \frac{\partial u}{\partial x} = D \frac{\partial^2 u}{\partial x^2}. \tag{10.0.4}$$

(Note that in the above limit $\alpha - \beta$ is of order of magnitude of Δx, i.e., $\alpha - \beta \sim \Delta x \frac{(\alpha+\beta)v}{2D}$.) Equation (10.0.4) holds for a concentration of chemical species which is both diffusing and drifting (with velocity v). If $\alpha > \beta$ then $v > 0$ and the drift is to the right.

Equation (10.0.4) also holds for the density of cells. But cells also react to stimulation by chemical signals, and move in the direction of the gradient of the signal intensity. This motion is called **chemotaxis**. To derive the chemotaxis equation we introduce the density of a chemical signal w and the strength of the signal $\gamma \tau(w)$. Then

$$T_i^{\pm} = \alpha + \gamma \left[\tau(w(x \pm \Delta x), t) - \tau(w(x, t)) \right] \tag{10.0.5}$$

Substituting this into the right-hand side of (10.0.1) and expanding by Taylor's formula, we get the limit equation, as $\Delta t \to 0$, $\Delta x \to 0$:

$$\frac{\partial u}{\partial t} = \frac{\partial}{\partial x}\left(D\frac{\partial u}{\partial x} - u\chi(w)\frac{\partial w}{\partial x}\right) \qquad (10.0.6)$$

where

$$\lim_{\Delta t \to 0} \alpha \frac{(\Delta x)^2}{\Delta t} = D, \quad D > 0.$$

$$2\gamma \frac{D}{\alpha}\frac{d\tau(w)}{dw} = \chi(w).$$

The function $\chi(w)$ is called the **chemotactic coefficient.** We can view Eq. (10.0.6) as a diffusion–advection equation with drift v given by

$$v = \chi(w)\frac{\partial w}{\partial x} = \frac{2\gamma D}{\alpha}\frac{d\tau}{dw}\frac{\partial w}{\partial x} = \frac{2\gamma D}{\alpha}\frac{\partial \tau(w)}{\partial x};$$

the velocity is positive in the direction of the gradient of the signal, i.e., $v > 0$ if $\partial\tau(w)/\partial x > 0$.

Proof of (10.0.6).

Let

$$u_i(t + \Delta t) - u_i(t) = T^+_{i-1}u_{i-1}(t) + T^-_{i+1}u_{i+1}(t) - (T^+_i + T^-_i)u_i(t)$$

where

$$T^+_i = \alpha + \gamma\left[\tau\left(w(x + \Delta x, t)\right) - \tau\left(w(x, t)\right)\right],$$
$$T^-_i = \alpha + \gamma\left[\tau\left(w(x - \Delta x, t)\right) - \tau\left(w(x, t)\right)\right].$$

Then

$$
\begin{aligned}
\frac{u(x, t + \Delta t) - u(x, t)}{\Delta t} =\ & \tfrac{1}{\Delta t}\left\{\alpha + \gamma\left[\tau\left(w(x, t)\right) - \tau\left(w(x - \Delta x, t)\right)\right]\right\} u(x - \Delta x, t) \\
+\ & \tfrac{1}{\Delta t}\left\{\alpha + \gamma\left[\tau\left(w(x, t)\right) - \tau\left(w(x + \Delta x, t)\right)\right]\right\} u(x + \Delta x, t) \\
-\ & \tfrac{1}{\Delta t}\left\{2\alpha + \gamma\left[\tau\left(w(x + \Delta x, t)\right) + \tau\left(w(x - \Delta x, t)\right)\right. \right. \\
& \left.\left. - 2\tau\left(w(x, t)\right)\right]\right\} u(x, t).
\end{aligned}
$$

$$(10.0.7)$$

We simplify notation:

$$
\begin{aligned}
w(x) &= w(x, t), & u(x) &= u(x, t), \\
w(x^+) &= w(x + \Delta x, t), & u(x^+) &= u(x + \Delta x, t), & (10.0.8)\\
w(x^-) &= w(x - \Delta x, t), & u(x^-) &= u(x - \Delta x, t).
\end{aligned}
$$

Then the right-hand side of (10.0.7) can be written as

$$
\begin{aligned}
& \frac{(\Delta x)^2}{\Delta t}\frac{\alpha\left[u(x^+) + u(x^-) - 2u(x)\right]}{(\Delta x)^2} + & & \gamma\frac{\Delta x}{\Delta t}\frac{\tau(w(x)) - \tau\left(w(x^+)\right)}{\Delta x}u(x^-) \\
& & & -\gamma\frac{\Delta x}{\Delta t}\frac{\tau\left(w(x^+)\right) - \tau(w(x))}{\Delta x}u(x^+) \\
& & & -\gamma\frac{(\Delta x)^2}{\Delta t}\frac{\tau\left(w(x^+)\right) + \tau\left(w(x^-)\right) - 2\tau(w(x))}{(\Delta x)^2}u(x) \\
& = & & J_1 + J_2 + J_3 + J_4.
\end{aligned}
$$

$$(10.0.9)$$

By Taylor's formula,

$$J_1 = \alpha \frac{(\Delta x)^2}{\Delta t} \frac{\partial^2 u(x)}{\partial x^2} + O\left(\frac{(\Delta x)^3}{\Delta t}\right), \tag{10.0.10}$$

$$J_3 = -\gamma \frac{\Delta x}{\Delta t} \frac{\partial \tau(x)}{\partial x} u(x^+) - \frac{1}{2}\gamma \frac{(\Delta x)^2}{\Delta t} \frac{\partial^2 \tau(x)}{\partial x^2} u(x^+) + O\left(\frac{(\Delta x)^3}{\Delta t}\right)$$

where $\tau(x) = \tau(w(x))$. Similarly,

$$J_2 = -\gamma \frac{\Delta x}{\Delta t} \frac{\partial \tau(x)}{\partial x} u(x^-) - \frac{1}{2}\gamma \frac{(\Delta x)^2}{\Delta t} \frac{\partial^2 \tau(x)}{\partial x^2} u(x^-) + O\left(\frac{(\Delta x)^3}{\Delta t}\right),$$

and

$$J_4 = -\gamma \frac{(\Delta x)^2}{\Delta t} \frac{\partial^2 \tau(x)}{\partial x^2} u(x) + O\left(\frac{(\Delta x)^3}{\Delta t}\right).$$

We next compute

$$
\begin{aligned}
J_3 + J_2 &= -\gamma \frac{(\Delta x)^2}{\Delta t} \frac{\partial \tau}{\partial x} \frac{u(x^+) - u(x^-)}{\Delta x} \\
&\quad - \gamma \frac{(\Delta x)^2}{\Delta t} \frac{\partial \tau}{\partial x} \frac{u(x) - u(x^-)}{\Delta x} \\
&\quad - \gamma \frac{(\Delta x)^2}{\Delta t} \frac{\partial^2 \tau}{\partial x^2} u(x) + O\left(\frac{(\Delta x)^3}{\Delta t}\right) \\
&= -2\gamma \frac{(\Delta x)^2}{\Delta t} \frac{\partial \tau}{\partial x} \frac{\partial u(x)}{\partial x} - \gamma \frac{(\Delta x)^2}{\Delta t} \frac{\partial^2 \tau}{\partial x^2} u(x) + +O\left(\frac{(\Delta x)^3}{\Delta t}\right)
\end{aligned}
$$

and

$$J_2 + J_3 + J_4 = -2\gamma \frac{(\Delta x)^2}{\Delta t} \frac{\partial \tau}{\partial x} \frac{\partial u}{\partial x} - 2\gamma \frac{(\Delta x)^2}{\Delta t} \frac{\partial^2 \tau}{\partial x^2} u(x) + O\left(\frac{(\Delta x)^3}{\Delta t}\right). \tag{10.0.11}$$

Now let $\Delta t \to 0$, so that $\alpha \frac{(\Delta x)^2}{\Delta t} \to D$. Then from (10.0.10), (10.0.11) we see that the right-hand side of (10.0.9), or (10.0.7), converges to

$$J \equiv D \frac{\partial^2 u}{\partial x^2} - \frac{2\gamma D}{\alpha}\left(\frac{\partial \tau}{\partial x} \frac{\partial u}{\partial x} + \frac{\partial^2 \tau}{\partial x^2} u(x)\right) \equiv D \frac{\partial^2 u}{\partial x^2} - J_0.$$

Setting

$$\chi(w) = \frac{2\gamma D}{\alpha} \frac{d\tau}{dw},$$

we also have

$$J_0 = \frac{2\gamma D}{\alpha}\left[\frac{\partial u}{\partial x}\left(\frac{\partial \tau}{\partial w} \frac{\partial w}{\partial x}\right) + u \frac{\partial}{\partial x}\left(\frac{\partial \tau}{\partial w} \frac{\partial w}{\partial x}\right)\right] = \frac{\partial}{\partial x}\left(u\chi(w) \frac{\partial w}{\partial x}\right).$$

Hence

$$\frac{u(x, t+\Delta t) - u(x, t)}{\Delta t} \to \frac{\partial}{\partial x}\left(D \frac{\partial u}{\partial x} - u\chi(w) \frac{\partial w}{\partial x}\right)$$

if $\Delta t \to 0$ and $\alpha \frac{(\Delta x)^2}{\Delta t} \to D$, and since the left-hand side of (10.0.7) also converges to $\partial u / \partial t$, the proof of (10.0.6) is complete.

PROBLEM 10.1. Determine how to choose the $T_i^{\pm}$ in order to obtain, by a random jump process, the diffusion equation

$$\frac{\partial u}{\partial t} = \frac{\partial}{\partial x}\left(D(x)\frac{\partial u}{\partial x}\right)$$

where $D(x)$ is a function of x. [Hint: Take $T_i^+ = \alpha(x_i + \frac{1}{2}\Delta x), T_i^- = \alpha(x_i - \frac{1}{2}\Delta x)$ where $\alpha(x)$ is a continuously differentiable function, $c = \lim_{\Delta t \to 0} \frac{(\Delta x)^2}{\Delta t}$, and $c\alpha(x) = D(x)$.]

PROBLEM 10.2. Bacteria with concentration $u(x,t)$ is chemoattracted to substance w, so that (10.0.6) holds. Assume that $\partial u / \partial x$, $\partial w / \partial x \to 0$ as $x \to \infty$. (i) Show that if $\chi(w) = const. = \chi$, then in steady state the bacteria density is related to the chemical substance $w(x)$ by

$$u(x) = const.\, e^{\chi w(x)/D}.$$

(ii) Compute, in steady state, $u(x)$, in terms of $w(x)$ when $\chi(w) = w^{\alpha}$ ($-\infty < \alpha < \infty$).

The diffusion equation (10.0.2), in n dimensions, has the form

$$\frac{\partial u}{\partial t} = \sum_{i,j=1}^{n} a_{ij}\frac{\partial^2 u}{\partial x_i \partial x_j} \tag{10.0.12}$$

where the matrix (a_{ij}) is positive definite, i.e., $\sum_{i,j=1}^{n} a_{ij}\xi_i\xi_j > 0$ if $\sum_{i=1}^{n} \xi_i^2 > 0$. This equation can be obtained by a random jumping process in n-dimensional lattice.

The diffusion equation with chemotaxis, (10.0.6), in n dimensions has the form

$$\frac{\partial u}{\partial t} = \sum_{i,j=1}^{n} a_{ij}\frac{\partial^2 u}{\partial x_i \partial x_j} - \nabla \cdot (u\chi(w)\nabla w).$$

The steady-state equation of (10.0.12) with a source $f(x)$ is the **elliptic** equation:

$$\sum_{i,j=1}^{n} a_{ij}\frac{\partial^2 u}{\partial x_i \partial x_j} = f(\mathbf{x}). \tag{10.0.13}$$

Consider the **Dirichlet problem** of solving (10.0.13) in a bounded domain Ω subject to a boundary condition

$$u = g \quad \text{on} \quad \partial\Omega. \tag{10.0.14}$$

This problem has a unique solution. Furthermore, the following **maximum principle** holds:

If $f \le 0$ and $g \ge 0$, then $u \ge 0$; moreover, either u is strictly positive in Ω or else $u \equiv 0$ in Ω.

There are corresponding results for the **parabolic** equation

$$\frac{\partial u}{\partial t} = \sum_{i,j=1}^{n} a_{ij} \frac{\partial^2 u}{\partial x_i \partial x_j} + f(\mathbf{x}, t) \tag{10.0.15}$$

for x in Ω and $t > 0$, with a boundary condition

$$u = g(\mathbf{x}, t) \quad \text{on} \quad \partial\Omega \times (0, \infty), \tag{10.0.16}$$

and an initial condition

$$u(\mathbf{x}, 0) = u_0(\mathbf{x}), \quad \mathbf{x} \in \Omega,$$

namely, a unique solution u exists, and if $f \le 0$, $g \ge 0$, $u_0 \ge 0$, then $u \ge 0$; for more details see [9].

10.1. Dictyostelium discoideum

A protozoa is a diverse group of unicellular eukaryotic microorganisms who get their food from the microenvironment; many of them are motile. The group includes a variety of parasites. Amoeba is a member of the protozoa genus and is commonly found in water and soil. It is surrounded by outer flexible membrane and it moves by means of pseudopodia. *Dictyostelium discoideum* is an amoeba residing in soil and commonly referred to as slime mold. The species undergoes a fascinating transition from unicellular amoebas into multicellular slug and then into a fruiting body. The lifecycle of *Dictyostelium discoideum* consists of four phases: vegetative, aggregation, migration, and culmination. The first phase begins with spore, a reproductive cell capable of developing into a new individual without fusion with another reproductive cell. In the vegetative phase the spore feeds on bacteria and grows into a mature cell. The second phase of aggregation begins when the amoebas run out of food. The sense of starvation initiates the creation of a machinery which secretes specific cyclic AMP (cAMP). This serves as stimulus to attract the amoebas by chemotaxis to the location of a central amoeba, the one dispensing the greatest amount of cAMP secretion. The amoebas begin to bump into each other and form a slug, a tight elongated mound of cells. The slug is 2–4 mm long, is composed of as many as 100,000 cells, and is capable of moving. Within the slug the amoebas differentiate into prestalk and prespore cells. During the migration phase the slug moves around until it finds a suitable environment to form a stalk of a fruiting body, with spores at the top. During the culmination phase the spores are dispersed by wind and germinate into a new generation of amoebas. Figure 10.1 is a schematic of the lifecycle of *Dictyostelium discoideum*.

Because of its simple lifecycle the *Dictyostelium discoideum* is commonly used as a model organism. It has been the subject of many studies, including some mathematical modeling. Here we focus on the aggregation phase where

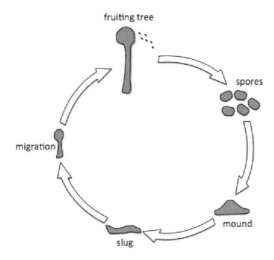

FIGURE 10.1. Schematics of the lifecycle of *Dictyostelium discoideum*

chemoattractic force pulls the amoebas toward one point, the location of the central amoeba. We take this point to be at the origin and introduce two variables:

$$n \;=\; \text{density of amoebas,}$$
$$A \;=\; \text{concentration of cAMP.}$$

Then, n satisfies the equation

$$\frac{\partial n}{\partial t} = D_n \nabla^2 n - \nabla \cdot (\chi n \nabla A), \tag{10.1.1}$$

where χ is a chemotactic parameter, and c satisfies the diffusion equation

$$\frac{\partial A}{\partial t} = D_A \nabla^2 A + \lambda n - \mu A; \tag{10.1.2}$$

here λ is the production rate of cAMP by the amoebas and μ is the degradation rate of cAMP. We assume spherical symmetry, so that the Laplacian $\nabla^2 u$ takes the form

$$\nabla^2 u = \frac{1}{r^2} \frac{\partial}{\partial r} \left(r^2 u \right)$$

where r is the distance to the origin.

We supplement the system with initial conditions

$$n(r,0) = 1, \quad A(r,0) = 0, \quad \text{for} \quad 0 < r < L \tag{10.1.3}$$

and boundary conditions

$$\frac{\partial n}{\partial r}(L,t) - \frac{1}{D_n} \chi n(L,t) \frac{\partial A}{\partial r}(L,t) = 0, \quad A(L,t) = 0, \quad \text{for} \quad t > 0. \tag{10.1.4}$$

10.2. Angiogenesis

Endothelial cells form the inner surface of blood vessels. Angiogenesis is a physiological process through which new blood vessels form from preexisting vessels. Angiogenesis occurs when there is an increased need for blood supply. This is commonly associated with cancer. In order to continue to grow abnormally beyond the diffusion-limited rate, tumor cells secrete **vascular endothelial growth factor** (VEGF), and these biochemical molecules attract endothelial cells from preexisting blood vessels to move toward the tumor, resulting in the formation of new vessels moving toward the tumor.

We model this process is terms of the following variables:

$$
\begin{aligned}
n &= \text{density of tumor cells,} \\
e &= \text{density of endothelial cells,} \\
h &= \text{concentration of VEGF.}
\end{aligned}
$$

For simplicity we assume that all the cells and VEGF move in one direction, the $x-$direction, and take

$$n = n(x,t), \quad h = h(x,t), \quad e = e(x,t).$$

The equation for the tumor cells is

$$\frac{\partial n}{\partial t} = D_n \frac{\partial^2 n}{\partial x^2} + \lambda\left(e\right) n \left(1 - \frac{n}{n_*}\right) - \mu_n n, \tag{10.2.1}$$

where μ_n is the death rate and where we assume that the growth rate $\lambda(e)$ is given by the Michaelis–Menten law:

$$\lambda(e) = \lambda_0 \frac{e}{1+e}, \quad \lambda_0 \text{ positive constant;} \tag{10.2.2}$$

thus without supply of blood (i.e., if $e = 0$) there is no growth. The VEGF concentration satisfies the equation

$$\frac{\partial h}{\partial t} = D_h \frac{\partial^2 h}{\partial x^2} + \lambda_h n - \mu_h h, \tag{10.2.3}$$

where λ_h is the production rate of h by tumor cells and μ_n is the degradation rate. Finally,

$$\frac{\partial e}{\partial t} = D_e \frac{\partial^2 e}{\partial x^2} - \frac{\partial}{\partial x}\left(\chi e \frac{\partial h}{\partial x}\right), \tag{10.2.4}$$

where χ is a chemotactic parameter.

We supplement the system (10.2.1)–(10.2.4) with initial conditions

$$n(x,0) = \left(L^2 - x^2\right) e^{-x^2}, \ h(x,0) = 0, \ e(x,0) = \gamma \frac{x^2}{L^2} e^{-(L-x)^2} \text{ for } 0 < x < L, \tag{10.2.5}$$

and boundary conditions

$$
\begin{aligned}
n_x(0,t) &= 0, & h_x(0,t) &= 0, & e_x(0,t) &= 0, \\
n(L,t) &= 0, & h_x(L,t) &= 0, & e(L,t) &= \gamma.
\end{aligned}
\tag{10.2.6}
$$

10.3. Diffusion/Dispersion in Population Dynamics

Most of the population dynamics models introduced in Chaps. 4 and 5 have meaningful biological counterparts when some of the populations undergo random walk which we model as diffusion; in this context one sometimes uses the word "dispersion" rather than "diffusion."

We give two examples. The first example is a SIR model where the infectious disease is rabies spread among foxes. While the healthy foxes stay in their territory, the rabid foxes are known to travel large distances, somewhat randomly, and attack other foxes. Since the disease is fatal, there is no chance for recovery, and the 2-dimensional model in the entire plane R^2 takes the form

$$
\frac{\partial S}{\partial t} = A - \beta IS - \mu S, \quad \text{in} \quad R^2,
\tag{10.3.1}
$$

$$
\frac{\partial I}{\partial t} = D\nabla^2 I + \beta IS - \gamma I, \quad \text{in} \quad R^2,
\tag{10.3.2}
$$

where μ is the natural death rate of healthy foxes and γ is the death rate of foxes infected with rabies. When $D = 0$, the DFE $(\frac{A}{\mu}, 0)$ is stable if

$$
A - \frac{\mu\gamma}{\beta} < 0, \quad \text{or} \quad \gamma > \frac{A\beta}{\mu},
\tag{10.3.3}
$$

that is, if the infected foxes die very quickly. The question we want to address is: What is the effect of dispersion when the DFE is stable, i.e., when (10.3.3) holds?

We assume that $S(x,0) \le \frac{A}{\mu}$ in R^2. Then from (10.3.1) it follows that $S(x,t) \le \frac{A}{\mu}$ in R^2 for all $t > 0$. Multiplying (10.3.2) by I and integrating over R^2, and assuming that $I(x,t)$ vanishes for all large $|x|$, we get

$$
\begin{aligned}
\frac{1}{2}\frac{d}{dt}\int_{R^2} I^2(x,t)dx &= -\int_{R^2} D|\nabla I|^2 dx + \int_{R^2}(\beta S - \gamma) I^2(x,t)dx \\
&\le \qquad\qquad -\varepsilon \int_{R^2} I^2(x,t)dx
\end{aligned}
$$

where we used the relations

$$
\beta S - \gamma \le \beta\frac{A}{\mu} - \gamma = -\varepsilon
$$

and $\varepsilon > 0$ by (10.3.3). Hence

$$
\int_{R^2} I^2(x,t)dx \le e^{-2\varepsilon t}\int_{R^2} I^2(x,0)dx \to 0 \quad \text{as} \quad t \to \infty.
$$

We conclude that any infection $I(x,0)$ will disappear as $t \to \infty$. Thus *dispersion helps eliminate any spatial infection.*

The second example deals with competition between two species, u and v. We take

$$\frac{\partial u}{\partial t} = D_1 u_{xx} + ru(1 - u) - uv, \qquad (10.3.4)$$

$$\frac{\partial v}{\partial t} = D_2 v_{xx} + rv(1 - v) - uv. \qquad (10.3.5)$$

If $D_1 = D_2 = 0$ then $(\frac{r}{1+r}, \frac{r}{1+r})$ is a steady point. Suppose $D_1 > 0$, $D_2 > 0$, $D_1 \neq D_2$ and the competition takes place in an interval $0 \leq x \leq L$, with

$$u(x, 0) = v(x, 0) = 1, \qquad (10.3.6)$$

$$u_x = v_x = 0 \quad \text{at} \quad x = 0 \quad \text{and} \quad u_x + u = v_x + v = 1 \quad \text{at} \quad x = L. \quad (10.3.7)$$

We shall say that u is the winner at time T if

$$\int_0^L u(x, T)dx > \int_0^L v(x, T)dx;$$

if the inequality is reversed, then we say that v is the winner at time T. An interesting question arises: Is the winner the species with the larger dispersion or with the smaller dispersion?

10.4. MATLAB Solver for Parabolic PDEs

The built-in MATLAB function "pdepe" solves the initial-boundary value problems for a system of parabolic PDEs in one dimension of the form

$$c\left(x, t, u, \frac{\partial u}{\partial x}\right)\frac{\partial u}{\partial t} = x^{-m}\frac{\partial}{\partial x}\left(x^m f\left(x, t, u, \frac{\partial u}{\partial x}\right)\right) + s\left(x, t, u, \frac{\partial u}{\partial x}\right)$$
$$(10.4.1)$$

for $t_0 \leq t \leq t_f$ and $a \leq x \leq b$. The interval $[a, b]$ must be finite; m is either 0, 1, or 2, corresponding to slab, cylindrical, or spherical symmetry, respectively. If $m > 0$, then a must be ≥ 0. The initial condition

$$u(x, t_0) = u_0(x)$$

needs to be specified (in the code) as a function. For all $t \geq 0$ and either $x = a$ or $x = b$, the solution components satisfy boundary conditions of the form

$$p(x, t, u) + q(x, t)f\left(x, t, u, \frac{\partial u}{\partial x}\right) = 0 \qquad (10.4.2)$$

at $x = a$ and $x = b$. Elements of q are either identically zero or never zero. Also, of the two coefficients, only p can depend on u. See MATLAB manual for more details.

In the call

sol = pdepe(m,pdefun,icfun,bcfun,xmesh,tspan) :

(1) m corresponds to m.
(2) pdefun computes the terms c, f, and s. It has the form
[c,f,s] = pdefun(x,t,u,dudx).
The input arguments are scalars x and t and vectors u and dudx that approximate the solution and its partial derivative with respect to

x, respectively. c, f, and s are column vectors. c stores the diagonal elements of the matrix (10.4.1).

(3) icfun evaluates the initial conditions. It has the form u = icfun(x). When called with an argument x, icfun evaluates and returns the initial values of the solution components at x in the column vector u.

(4) bcfun evaluates the terms and of the boundary conditions (10.4.2). It has the form

[pl,ql,pr,qr] = bcfun(xl,ul,xr,ur,t)

. ul is the approximate solution at the left boundary xl =a, and ur is the approximate solution at the right boundary xr =b. pl and ql are column vectors corresponding to p and q evaluated at xl, and similarly pr and qr correspond to xr.

(5) xmesh(1) and xmesh(end) correspond to a and b.

(6) tspan(1) and tspan(end) correspond to t_0 and t_f.

In Algorithm 16 and 17, we demonstrate how to use pdepe to solve a scalar and a system of parabolic equations, respectively.

(1) The scalar equation we solve here is

$$\frac{\partial u}{\partial t} = \frac{1}{\pi^2} \frac{\partial}{\partial x} \left(\frac{\partial u}{\partial x} \right), \tag{10.4.3}$$

on an interval $0 \le x \le 1$ for times $t \ge 0$, with the initial condition

$$u(x,0) = \sin(\pi x),$$

and boundary conditions

$$\begin{cases} \frac{\partial u}{\partial x}(1,t) &= \pi e^{-t}, \\ u(1,t) &= 0. \end{cases} \tag{10.4.4}$$

The exact solution of this scalar equation is

$$u(x,t) = e^{-t} \sin(\pi x).$$

Note that we must rewrite the Eq. (10.4.3) and the boundary conditions (10.4.4) in the matching form with (10.4.1) and (10.4.2), i.e.,

$$\pi^2 \frac{\partial u}{\partial t} = \frac{\partial}{\partial x} \left(\frac{\partial u}{\partial x} \right), \tag{10.4.5}$$

and

$$\begin{cases} -\pi e^{-t} + \frac{\partial u}{\partial x}(1,t) &= 0, \\ u(1,t) &= 0. \end{cases} \tag{10.4.6}$$

Thus the choices of m, c, f, and s are

$$m = 0, \quad c = \pi^2, \quad f = \frac{\partial u}{\partial x}, \quad s = 0,$$

and the choices of p and q are

$$p(0, t, u) = \text{pl} = -\pi e^{-t}, \quad q(0, t, u) = \text{ql} = 1,$$

$$p(1, t, u) = \text{ur} = u(1, t), \quad q(1, t, u) = \text{qr} = 0.$$

The code is demonstrated in Algorithm 16 with 20 grids in space and the final time $t = 1$. The solution $u(x, t)$ is shown in Fig. 10.2a with respect to x and t. The solution at the final time $u(x, 1)$ is shown in Fig. 10.2b. We can see that the solution becomes flatter as time goes on.

Algorithm 16 pdepe_ex1.m
___

```
function pdepe_ex1
m = 0;
x = linspace(0,1,20);
t = linspace(0,1,10);
sol = pdepe(m,@pdepe_ex1pde,@pdepe_ex1ic,@pdepe_ex1bc,x,t);
% Extract the first solution component as u.
u = sol(:,:,1);
% The surface plot of the solution u(x,t)
surf(x,t,u)
title('Numerical solution computed with 20 mesh points.')
xlabel('x')
ylabel('t')
% The figure consists of the solution u(x,1) in blue line and the exact
solution
% in red dot
figure plot(x,u(end,:),'b',x,exp(-1)*sin(pi*x),'ro')
xlabel('x')
ylabel('u(x,1)')
legend('numerical solution','exact solution')
% --------------------------------------------------
function [c,f,s] = pdepe_ex1pde(x,t,u,DuDx)
c = pi^2; f = DuDx; s = 0;
% --------------------------------------------------
function u0 = pdepe_ex1ic(x)
u0 = sin(pi*x);
% --------------------------------------------------
function [pl,ql,pr,qr] = pdepe_ex1bc(xl,ul,xr,ur,t)
pl = -pi * exp(-t);
ql = 1;
pr = ur;
qr = 0;
```
___

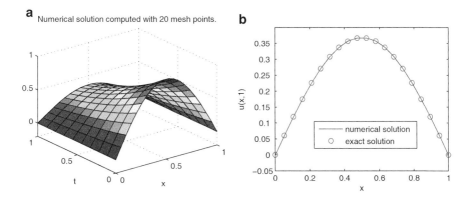

FIGURE 10.2. (a) The solution $u(x,t)$ of the scalar equation (10.4.3) (b). The solution u at $t = 1$

(2) Here we demonstrate how to use pdepe to solve the system of parabolic equations (10.3.4)–(10.3.7). Comparing with the standard forms (10.4.1) and (10.4.2) in MATLAB built-in function "pdepe", we have $m = 0$:

$$c = \begin{bmatrix} \frac{1}{D_1} \\ \frac{1}{D_2} \end{bmatrix}, \quad f = \begin{bmatrix} \frac{\partial u}{\partial x} \\ \frac{\partial v}{\partial x} \end{bmatrix}, \quad s = \begin{bmatrix} \frac{ru(1-u)-uv}{D_1} \\ \frac{rv(1-v)-uv}{D_2} \end{bmatrix},$$

$$p(0,t,u) = \mathrm{pl} = \begin{bmatrix} 0 \\ 0 \end{bmatrix}, \quad q(0,t,u) = \mathrm{ql} = \begin{bmatrix} 1 \\ 1 \end{bmatrix},$$

$$p(L,t,u) = \mathrm{pr} = \begin{bmatrix} u(L,t)-1 \\ v(L,t)-1 \end{bmatrix} = \mathrm{ur\text{-}1}, \quad \text{and} \quad q(L,t,u) = \mathrm{qr} = \begin{bmatrix} 1 \\ 1 \end{bmatrix}.$$

The code is demonstrated in Algorithm 17 with 20 grids in space and the final time $t = 1$. The solutions $u(x,t)$ and $v(x,t)$ are shown in Fig. 10.3 with respect to x and t. We see that the solutions satisfy $u(x,t) < v(x,t)$.

PROBLEM 10.3. Give an answer to the question in the second example in Sect. 10.3 in case $r = 1$, $L = 3$, $D_1 = 1$, $D_2 = 2$ for $T = 5$, and $T = 10$. Is the winner the species with a larger dispersion or with a smaller dispersion?

Now we explain how to use pdepe to solve the radially symmetric case of (10.1.1) and (10.1.2) which satisfies

$$\frac{1}{D_n}\frac{\partial n}{\partial t} = r^{-2}\frac{\partial}{\partial r}\left(r^2\left(\frac{\partial n}{\partial r} - \frac{1}{D_n}\chi n\frac{\partial A}{\partial r}\right)\right), \tag{10.4.7}$$

and

$$\frac{1}{D_A}\frac{\partial A}{\partial t} = \frac{1}{r^2}\frac{\partial}{\partial r}\left(r^2\frac{\partial A}{\partial r}\right) + \frac{1}{D_A}\left(\lambda n - \mu A\right), \tag{10.4.8}$$

Algorithm 17 pdepe_ex2.m

```
function pdepe_ex2
m = 0;
L = 3;
T = 1;
x = linspace(0,L,20);
t = linspace(0,T,10);
sol = pdepe(m,@pdepe_ex2pde,@pdepe_ex2ic,@pdepe_ex2bc,x,t);
u = sol(:,:,1);
v = sol(:,:,2);
figure surf(x,t,u); box on
title('u(x,t)')
xlabel('x')
ylabel('t')
figure
surf(x,t,v); box on
title('v(x,t)')
xlabel('x')
ylabel('t')
% ─────────────────────────────────────
function [c,f,s] = pdepe_ex2pde(x,t,u,DuDx)
D1 = 1; D2 = 2;
r = 1;
c = [1/D1; 1/D2];
f = [1; 1] .* DuDx;
s = [(r*u(1)*(1-u(1))-u(1)*u(2))/D1; (r*u(2)*(1-u(2))-u(1)*u(2))/D2];
% ─────────────────────────────────────
function u0 = pdepe_ex2ic(x);
u0 = [1; 1];
% ─────────────────────────────────────
function [pl,ql,pr,qr] = pdepe_ex2bc(xl,ul,xr,ur,t)
pl = [0; 0];
ql = [1; 1];
pr = [ur(1)-1; ur(2)-1];
qr = [1; 1];
```

---

with initial conditions

$$n(r,0) = 1, \quad A(r,0) = 0, \quad \text{for} \quad 0 < r < L, \qquad (10.4.9)$$

and boundary conditions

$$\frac{\partial n}{\partial r}(0,t) = \frac{\partial n}{\partial r}(L,t) - \frac{1}{D_n}\chi n(L,t)\frac{\partial A}{\partial r}(L,t) = 0, \quad \frac{\partial A(0,t)}{\partial r} = A(L,t) = 0,$$

$$\text{for} \quad t > 0. \qquad (10.4.10)$$

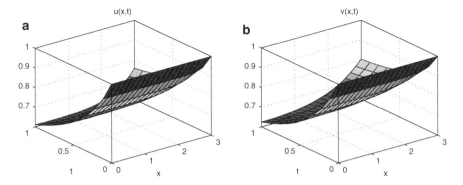

FIGURE 10.3. The solution of the system of parabolic equations (10.3.4)–(10.3.7); (a) $u(x,t)$ (b) $v(x,t)$

Comparing with the standard forms (10.4.1) and (10.4.2) in MATLAB built-in function pdepe, we have $m = 2$:

$$c = \begin{bmatrix} \frac{1}{D_n} \\ \frac{1}{D_c} \end{bmatrix}, \quad f = \begin{bmatrix} \frac{\partial n}{\partial r} - \frac{1}{D_n}\chi n \frac{\partial A}{\partial r} \\ \frac{\partial A}{\partial r} \end{bmatrix}, \quad s = \begin{bmatrix} 0 \\ \frac{1}{D_A}(\lambda n - \mu A) \end{bmatrix},$$

and

$$p(0,t,u) = \mathrm{pl} = \begin{bmatrix} 0 \\ 0 \end{bmatrix}, \quad q(0,t,u) = \mathrm{ql} = \begin{bmatrix} 1 \\ 1 \end{bmatrix},$$

$$p(L,t,u) = \mathrm{pr} = \begin{bmatrix} 0 \\ A(L,t) \end{bmatrix}, \quad q(L,t,u) = \mathrm{qr} = \begin{bmatrix} 1 \\ 0 \end{bmatrix}.$$

PROBLEM 10.4. Take nondimensional parameters $D_n = 10^{-3}$, $D_c = 0.5$, $\chi = 0.1$, $\lambda = 4$, $\mu = 20$, and $L = 3$, and use pdepe to compute the profiles of $n(r,t)$ and $c(r,t)$ in (10.1.1)–(10.1.4) for $t = 1, 5, 10$.

PROBLEM 10.5. Take nondimensional parameters $D_n = 10^{-4}$, $D_h = 0.5$, $D_e = 10^{-5}$, $\lambda_0 = 2.6$, $\mu_n = 1.2$, $n_* = 1$, $\lambda_h = 4$, $\mu_h = 30$, $\chi = 0.06$, $L = 5$, and $\gamma = 2$. Use pdepe to compute the profiles of $n(x,t)$, $h(x,t)$, and $e(x,t)$ in (10.2.1)–(10.2.6) at $t = 1, 5, 10$.

PROBLEM 10.6. A population of cells with density $p(x,t)$ satisfies the diffusion equation

$$\frac{\partial p}{\partial t} = Dp_{xx} + f(c)p \quad \text{in} \quad -1 < x < 1, t > 0$$

where c is the concentration of oxygen, and

$$\frac{\partial c}{\partial t} = c_{xx} - 3cp \quad \text{in} \quad -1 < x < 1, t > 0;$$

$3cp$ is the consumption rate by the cells. Boundary and initial conditions are

$$p(\pm 1, t) = 1, \quad c(\pm 1, t) = 4,$$
$$p(x,0) = 1, \quad c(x,0) = 4.$$

The function $f(c)$ is given by

$$f(x) = \begin{cases} 0 & \text{if} \quad c < 1 \text{ or } c > 4 \\ c-1 & \text{if} \quad 1 < c < 2 \\ 1 & \text{if} \quad 2 < c < 3 \\ 4-c & \text{if} \quad 3 < c < 4, \end{cases}$$

reflecting the fact that too little oxygen or too much oxygen are harmful to the cells. Compute the total population of cells:

$$P = P(t, D) = \int_{-1}^{1} p(x, t) dx$$

as a function of t for several different values of D. Can you draw any conclusions on the dependence of P on D?

CHAPTER 11

Cancer

An abnormally new growth of tissue that grows more rapidly than normal cells and has no physiological function is called **neoplasm**, or **tumor**. Such growth competes with normal cells for space and nutrients. When this new growth is localized, it is called **benign tumor**, while if it spreads to other parts of the body, it is called **malignant tumor** or **cancer**, although sometimes people use the words tumor and cancer interchangeably.

When tumor in a tissue has reached a size of several millimeters, it may no longer grow because of limited diffusion of nutrients from the vascular system. Such a tumor is called **avascular**. Tumor cells will try to induce the formation of new blood vessels (**angiogenesis**) and direct their movement toward the tumor. They do so by secreting **tumor endothelial growth factor** (VEGF) and, if successful, the tumor becomes **vascular**. A vascular tumor continues to grow and may invade the lymphatic or blood vessels. Transported by the fluid flow in these vessels, cancer cells may spread into other locations in the body, resulting in **metastasis**, that is, the formation of new cancer colonies. Most cancer deaths are due to metastasized cancer.

Cancer is a disease of tissue growth failure, and it is the result of normal cells transforming into cancer cells because of mutations in genes that regulate cell growth and differentiation. In the context of cancer, these genes are classified into **oncogenes** and **tumor suppressor genes**. Oncogenes are genes which promote cell growth and reproduction. Tumor suppressor genes are genes which inhibit cell division and survival. Malignant transformation occurs when oncogenes become overexpressed compared to normal oncogenes or when tumor suppressor genes become underexpressed or disabled. Typically several gene mutations are required in order to transform a normal cell into a cancer cell.

It is commonly believed that most mutations leading to cancer are due to external conditions, such as smoking, dietary factors, environmental pollutants, exposure to radiation, and certain infections. But some mutations are hereditary.

There are more than one hundred known types of human cancer, broadly categorized according to the tissue of origin. **Carcinomas** begin with epithelial cells; **sarcomas** arise from connective tissues, muscles, and vasculature;

© Springer International Publishing Switzerland 2014
A. Friedman, C.-Y. Kao, *Mathematical Modeling of Biological Processes*, Lecture Notes on Mathematical Modelling in the Life Sciences, DOI 10.1007/978-3-319-08314-8_11

leukemias and **lymphomas** are cancers of the hematopoietic (blood) and immune system, respectively; **gliomas** are cancers of the central nervous system, including the brain; **retinoblastomas** are cancers of the eyes.

Cancer is a complex disease involving several scales of space and time. The spatial scales include the molecular scale within a single cancer cell and the macroscopic scale of densities of cell populations growing in a tissue. The temporal scales are the cell lifecycle scale, typically on order of days, and the scale of months or years it takes for cancer to become diagnosed. There have been many mathematical models that try to capture some aspects of cancer growth; some of the models are of phenomenological character while others are tailored to specific experiments. Both types of models continue to add to our understanding of the disease. In this chapter we introduce two phenomenological generic models of tumor growth. The first one assumes that the tumor tissue consists only of live tumor cells, uniformly distributed. The second model includes also dead cells.

Model 1. Only Live Cancer Cells

We assume that the tumor consists only of uniformly distributed proliferating cancer cells. We denote by c the concentration of nutrients in the tumor and assume that the proliferation rate is given by the rule

$$\text{proliferation rate} = \mu(c - \tilde{c})$$

where μ and $\tilde{c}$ are positive constants. Thus if $c > \tilde{c}$ then the tumor grows while if $c < \tilde{c}$ then the tumor shrinks. Since the cells are uniformly distributed with fixed density, proliferation induces movement of cells, and we denote the resulting velocity by $\mathbf{v}$. We next assume that the tissue has the consistency of a porous medium. Hence, by Darcy's law and nondimensionalization:

$$\mathbf{v} = -\nabla \sigma$$

where σ is the pressure of the moving cells within the tumor.

By conservation of mass we also have

$$\text{div} \mathbf{v} = \mu (c - \tilde{c}) .$$

If we denote by $\Omega(t)$ the tumor region at time t, we then obtain the following equation for σ:

$$\nabla^2 \sigma = -\mu (c - \tilde{c}) \quad \text{in} \quad \Omega(t), t > 0. \tag{11.0.1}$$

For avascular tumor the nutrient concentration satisfies the diffusion equation, herein written in nondimensionalized form:

$$\frac{\partial c}{\partial t} - \nabla^2 c + c = 0 \quad \text{in} \quad \Omega(t), t > 0 \tag{11.0.2}$$

where the term c represents consumption of nutrients by the uniformly distributed cancer cells.

We introduce boundary conditions

$$c = \tilde{c} \quad \text{on the boundary } \partial\Omega(t), t > 0, \tag{11.0.3}$$

$$\sigma = \gamma \kappa \quad \text{on} \quad \partial\Omega(t), t > 0, \tag{11.0.4}$$

where $\bar{c} > \tilde{c}$, γ is a positive parameter, and κ is the mean curvature of $\partial\Omega(t)$: the curvature is the a result of cells' adhesion at the tumor boundary. For a sphere, κ is the reciprocal of the radius.

We assume that the boundary points on $\partial\Omega(t)$ are moving with normal velocity v_n, that is, the velocity $\mathbf{v}$ in the outward normal direction, so that

$$v_n = \mathbf{v} \cdot \mathbf{n} = -\frac{\partial\sigma}{\partial n} \quad \text{on} \quad \partial\Omega(t), t > 0. \tag{11.0.5}$$

We supplement the system $(11.0.1)$–$(11.0.5)$ with initial conditions

$$\begin{aligned} \Omega(0) &= \Omega_0, \\ c(x,0) &= c_0(x) \ge 0 \quad \text{for} \quad x \in \Omega_0. \end{aligned} \tag{11.0.6}$$

Notice that the above equations are nondimensionalized except for the growth rate μ and the surface-tension coefficient γ.

We next specialize to radially symmetric solutions with $\Omega(t) = \{r < R(t)\}$. By integrating both sides of $(11.0.1)$ and using $(11.0.5)$, one obtains

$$R^2(t)\frac{dR(t)}{dt} = \int_0^{R(t)} \mu(c(r,t) - \bar{c})r^2 dr \tag{11.0.7}$$

A proof that there exists a unique radially symmetric solution for all $t > 0$ is given in [10], where the existence of a unique radially symmetric stationary solution is also proved.

PROBLEM 11.1. Show that a radially symmetric stationary solution of $(11.0.1)$–$(11.0.5)$ must be given by

$$c_s(r) = \bar{c}\frac{R_s}{\sinh R_s}\frac{\sinh r}{r}, \quad \sigma_s(r) = C - \mu c_s(r) + \frac{\mu}{6}\tilde{c}r^2,$$

where

$$C = \frac{\gamma}{R_s} + \mu\bar{c} - \frac{\mu\tilde{c}R_s^2}{6},$$

and R_s is a solution to the equation

$$\tanh R_s = \frac{R_s}{1 + \left(\frac{\tilde{c}}{3\bar{c}}\right)R_s^2}. \tag{11.0.8}$$

PROBLEM 11.2. Prove that Eq. $(11.0.8)$ has a unique positive solution. [Hint: Setting $h(x) = \frac{1}{x^2} + \Lambda - \frac{1}{x}\coth(x)$ $(\Lambda = \frac{1}{3}\frac{\tilde{c}}{\bar{c}})$, you need to prove that $h(x)$ has a unique zero. Since $h(x) \to \Lambda - \frac{1}{3} < 0$ if $x \to 0$, $h(x) \to \Lambda$ if $x \to \infty$, it suffices to show that $h(x)$ is monotone increasing. Compute

$$h'(x) = \frac{k(x)}{x^3(\sinh(x))^2}$$

and prove that $k(x) > 0$ by showing that $k^{(j)}(0)=0$ $(0 \le j \le 3)$, $k^{(4)}(x) = 16x\cosh(x)\sinh(x)$.]

Model 2. Live and Dead Cancer Cells

In this model we include dead cancer cells. Cell death is a process that can occur in two ways:

(i) **Apoptosis**, where the cell participates in the process, the nucleus begins to break apart, and the cell breaks into several pieces in orderly fashion

(ii) **Necrosis**, where the death occurs in response to adverse conditions in cell's environment and the cell membrane ruptures and releases cell's contents

Apoptosis is a clean programmed death, whereas necrosis is a messy death.

When avascular tumor grows to several millimeters the cells in the core do not receive enough nutrients (e.g., oxygen) from the vasculature, and they undergo necrosis.

We denote by c_0 the critical level of nutrient concentration below which the cell undergoes necrosis. Then the density of live cancer cells satisfies the equation

$$\frac{\partial n}{\partial t} + \operatorname{div}(n\mathbf{v}) = \lambda \frac{(c - c_0)^+ n}{K + c} - \mu (c_0 - c)^+ n \qquad (11.0.9)$$

where $x^+ = x$ if $x > 0$, $x^+ = 0$ if $x \le 0$. The first term on the right-hand side is a growth term modeled by the Michaelis–Menten law, and the last term represents death by necrosis.

The density b of dead cells satisfies the equation

$$\frac{\partial b}{\partial t} + \operatorname{div}(b\mathbf{v}) = \mu (c_0 - c)^+ n - \gamma b \qquad (11.0.10)$$

where γ is the removal rate of the dead cells (or their debris). Next we write the equation for nutrient concentration c:

$$\frac{\partial c}{\partial t} - D_c \nabla^2 c + \bar{\lambda} \frac{(c - c_0)^+ n}{K + c} = \alpha \qquad (11.0.11)$$

where $\lambda/\bar{\lambda}$ is the gain of biomass from nutrients, and

$$\alpha = 0 \qquad \text{for avascular tumor,}$$
$$\alpha > 0 \qquad \text{for vascular tumor.}$$

We shall consider only the radially symmetric case with $r = |x|$, so that $n = n(r,t)$, $b = b(r,t)$, $c = c(r,t)$, and, for any function $m(r,t)$,

$$\operatorname{div}(m\mathbf{v}) = \frac{1}{r^2} \frac{\partial}{\partial r} (r^2 m v) \quad \text{where} \quad \mathbf{v} = \frac{x}{r} v, \quad v = v(r,t).$$

We assume that the combined live and dead cells are uniformly distributed, so that, in appropriate units,

$$n + b \equiv 1. \qquad (11.0.12)$$

By adding Eq. (11.0.9) and (11.0.10) we then get

$$\frac{1}{r^2}\frac{\partial}{\partial r}\left(r^2 v\right) = \lambda\frac{(c-c_0)^+ n}{K+c} - \gamma b,$$

or

$$v(r,t) = \frac{1}{r^2}\int_0^r s^2\left[\lambda\frac{(c-c_0)^+ n}{K+c} - \gamma(1-n)\right]ds. \tag{11.0.13}$$

Equations (11.0.9)–(11.0.13) hold in the tumor region $\{r < R(t)\}$. We take the boundary conditions

$$n = 1, b = 0, c = \bar{c} \quad \text{on} \quad r = R(t), t > 0 \tag{11.0.14}$$

with $\bar{c} > c_0$ and prescribe initial conditions

$$n(r,0) = 1, b(r,0) = 0, c(r,0) = \bar{c} \quad \text{for} \quad 0 \le r \le R(0). \tag{11.0.15}$$

Notice that in (11.0.14) we assumed that all the cells at the tumor boundary are live cells. We expect that the density of $n(r,t)$ will be an increasing function in r for any $t > 0$.

PROBLEM 11.3. Take $\lambda = 2.5$, $c_0 = 0.1$, $K = 1$, $\mu = 1.3$, $\gamma = 0.5$, $D_c = 0.5$, $\bar{\lambda} = 4$, and $R(0) = 1$. Compute and simulate the profile of $n(r,t)$ at $t = 10$ for $\alpha = 0$ (avascular tumor) and $\alpha = 0.8$ (vascular tumor).

11.1. Numerical Approach for the Cancer Cells Model

Here we show how to solve the tumor model of the system (11.0.1)–(11.0.5) in the radially symmetric case. In order to do this, we first transform the Cartesian coordinate system to the spherical coordinate system:

$$\begin{aligned} x &= r\sin(\phi)\cos(\theta), \\ y &= r\sin(\phi)\sin(\theta), \\ z &= r\cos(\phi). \end{aligned}$$

In the spherical coordinate system, the Laplace operator is

$$\nabla^2 = \frac{1}{r^2}\frac{\partial}{\partial r}\left(r^2\frac{\partial}{\partial r}\right) + \frac{1}{r^2\sin\phi}\frac{\partial}{\partial\phi}\left(\sin\phi\frac{\partial}{\partial\phi}\right) + \frac{1}{r^2\sin^2(\phi)}\frac{\partial^2}{\partial\theta^2}.$$

Thus, the system of Eqs. (11.0.1)–(11.0.5) in the radially symmetric case becomes

$$\frac{\partial c}{\partial t} - \left(\frac{\partial^2}{\partial r^2} + \frac{2}{r}\frac{\partial}{\partial r}\right)c + c = 0 \quad \text{in } \Omega(t),\, t > 0, \tag{11.1.1}$$

$$\left(\frac{\partial^2}{\partial r^2} + \frac{2}{r}\frac{\partial}{\partial r}\right)\sigma = -\mu(c - \tilde{c}) \quad \text{in } \Omega(t),\, t > 0, \tag{11.1.2}$$

$$c = \bar{c} \quad \text{on } \partial\Omega(t), t > 0, \text{ and } \bar{c} > \tilde{c}, \tag{11.1.3}$$

$$\sigma = \gamma\frac{1}{R(t)} \quad \text{on } \partial\Omega(t),\, t > 0, \tag{11.1.4}$$

$$\frac{\partial\sigma}{\partial r} = -v_n \quad \text{on } \partial\Omega(t), t > 0. \tag{11.1.5}$$

Note that the curvature is given by $\kappa = \frac{1}{R(t)}$ at all the points on the boundary $\partial\Omega(t)$.

The presence of a moving boundary $r = R(t)$ makes direct design of numerical methods difficult. A straightforward way to handle the moving boundary problem in the radially symmetric case is to map the moving boundary into a fixed one by introducing a new coordinate:

$$\rho = \frac{r}{R(t)} \quad T = t.$$

Then

$$\frac{\partial}{\partial t} = \frac{\partial}{\partial T} - \frac{rR'(T)}{R^2(T)}\frac{\partial}{\partial\rho}, \quad \frac{\partial}{\partial r} = \frac{1}{R(T)}\frac{\partial}{\partial\rho}.$$

Setting $\Omega_0 = \{r < 1\}$, the system (11.1.1)–(11.1.5) becomes

$$\frac{\partial c}{\partial t} - \left(\frac{\partial^2}{\partial r^2} + \frac{2}{r}\frac{\partial}{\partial r}\right)c + c = 0 \quad \text{in } \Omega_0, \, t > 0, \tag{11.1.6}$$

$$\left(\frac{\partial^2}{\partial r^2} + \frac{2}{r}\frac{\partial}{\partial r}\right)\sigma = -\mu(c - \tilde{c}) \quad \text{in } \Omega_0, \, t > 0, \tag{11.1.7}$$

$$c = \bar{c} \quad \text{on } r = 1, t > 0, \text{ and } \bar{c} > \tilde{c}, \tag{11.1.8}$$

$$\sigma = \gamma\frac{1}{R(t)} \quad \text{on } r = 1, t > 0, \tag{11.1.9}$$

$$\frac{\partial\sigma}{\partial r} = -\frac{dR}{dt} \quad \text{on } r = 1, t > 0. \tag{11.1.10}$$

The solution of Eq. (11.1.7) can be solved exactly and is

$$\sigma(t, r) = -\frac{\mu(c(t, r) - \tilde{c})r^2}{6} + \frac{\mu(c(t, r) - \tilde{c})}{6} + \gamma\frac{1}{R(t)}.$$

From Eq. (11.1.10), the rate of change in radius $R(t)$ satisfies

$$\frac{dR(t)}{dt} = \frac{\mu(c(t, r) - \tilde{c})}{3}. \tag{11.1.11}$$

We can then solve the system of one parabolic equation (11.1.6) with an initial condition and boundary conditions (11.1.8) and $\partial c/\partial r(0) = 0$ and an ordinary differential equation (11.1.11) with an initial condition.

PROBLEM. 10.3. Write a code to solve the aforementioned system by using MATLAB built-in function "pdepe" which was introduced in Sect. 10.4 for Eq. (11.1.6) and Euler method for the ordinary differential equation (11.1.11).

PROBLEM 11.4. Draw the graph of the stationary radius R_s as a function of $\tilde{c}/3\bar{c}$.

PROBLEM 11.5. Take a small radially symmetric perturbation of the stationary solution with radius R_s:

$$\Omega_0 = \{r < R_s\}, \quad c_0(r) = c_s(r) + \varepsilon, \quad \varepsilon = \frac{1}{100}.$$

Compute the tumor's boundary $R(t)$ of the system (11.1.6)–(11.1.10) for $\gamma = 1$ and

$$\mu = 0.1, 1, 10, 100, 1000.$$

In which cases does $R(t) \to R_s$ as t increases to infinity? What is the biological implication?

CHAPTER 12

Cancer Therapy

Cancer treatment consists of surgery, chemotherapy, and radiation. Surgery is the primary method of treatment. In localized cancer, surgery typically attempts to remove the entire mass and, in some cases, the nearby lymph nodes. Chemotherapy, in addition to surgery, has proven useful in different types of cancer such as breast cancer, colorectal cancer, and ovarian cancer. Radiation is used in addition to surgery and chemotherapy, for example, in painful bone metastasis.

The scheduling of chemotherapy is done on ad hoc basis, and it is not at all clear whether or how to determine a schedule of treatment that will yield the best results. Mathematical models that are validated by experimental results can play a useful role as tools to evaluate the results of different chemotherapeutic protocols of treatment and suggest how best to schedule them.

In this chapter we develop two completely different treatments of cancer. The first one is by oncolytic virus, and the second one is by injection of drug GM-CSF into the tumor.

12.1. Viral Therapy

Viral therapy is a therapy which employs oncolytic virus. Oncolytic viruses are genetically altered viruses that can infect and reproduce in cancer cells, but leave healthy normal cells unharmed. They are currently being developed as a potential drug for cancer treatment. When injected into the cancer tissue, they invade cancer cells. Upon lysis of an infected cell, a swarm of new viruses burst out of the dead cell and infect neighboring tumor cells. Tumor therapy by oncolytic viruses has been and continues to be actively tested in clinical trials for a variety of malignancies.

At present, despite producing hundreds of infectious viruses per one infected tumor cell, most virus species are unable to eradicate the tumor. There is increasing evidence that the host response to an active viral infection plays a critical role in determining the overall efficacy of viral therapy. Indeed, it has been demonstrated that the innate immune system destroys infected cells (as well as free virus particles), thus enabling the tumor to grow.

© Springer International Publishing Switzerland 2014
A. Friedman, C.-Y. Kao, *Mathematical Modeling of Biological Processes*, Lecture Notes on Mathematical Modelling in the Life Sciences, DOI 10.1007/978-3-319-08314-8_12

To mathematically model the process of oncolytic viral therapy we consider a very simple model of cancer and introduce the following variables:

$$
\begin{aligned}
x &= \text{number density of cancer cells,} \\
y &= \text{number density of infected cancer cells,} \\
n &= \text{number density of dead cells,} \\
v &= \text{number density of free viruses (those} \\
  &\quad\ \text{which are not contained in cancer cells).}
\end{aligned}
$$

We assume that the tumor is radially symmetric and denote by $u(r,t)$ the radial velocity, $u(0,t) = 0$; then, by conservation of number density, the following equations hold in the tumor region $\{r < R(t)\}$:

$$
\frac{\partial x}{\partial t} + \frac{1}{r^2}\frac{\partial}{\partial r}\left(r^2 ux\right) = \lambda x - \beta xv, \tag{12.1.1}
$$

$$
\frac{\partial y}{\partial t} + \frac{1}{r^2}\frac{\partial}{\partial r}\left(r^2 uy\right) = \beta xv - \delta y, \tag{12.1.2}
$$

$$
\frac{\partial n}{\partial t} + \frac{1}{r^2}\frac{\partial}{\partial r}\left(r^2 un\right) = \delta y - \mu n, \tag{12.1.3}
$$

$$
\frac{\partial v}{\partial t} - D\frac{1}{r^2}\frac{\partial}{\partial r}\left(r^2 \frac{\partial v}{\partial r}\right) = b\delta y - \gamma v. \tag{12.1.4}
$$

Here

$$
\begin{aligned}
\lambda &= \text{proliferation rate of tumor cells,} \\
\beta &= \text{infection rate,} \\
\delta &= \text{infected-cell lysis rate,} \\
\mu &= \text{removal rate of dead cells,} \\
D &= \text{diffusion coefficient of viruses,} \\
b &= \text{burst size of infected cells,} \\
\gamma &= \text{clearance rate of virus.}
\end{aligned}
$$

We assume that

$$
x(r,t) + y(r,t) + n(r,t) = \theta, \quad \theta \text{ is constant.} \tag{12.1.5}
$$

Taking the sum of Eqs. (12.1.1)–(12.1.3) we obtain

$$
\frac{\theta}{r^2}\frac{\partial}{\partial r}\left(r^2 u\right) = \lambda x - \mu n,
$$

or

$$
\frac{\theta}{r^2}\frac{\partial}{\partial r}\left(r^2 u\right) = \lambda x - \mu(\theta - x - y), \tag{12.1.6}
$$

and we may then replace Eq. (12.1.3) by (12.1.6).

To complete the mathematical model, we prescribe boundary conditions

$$
\frac{\partial v}{\partial r}(R(t),t) = 0, \quad t > 0, \tag{12.1.7}
$$

$$
\dot{R}(t) = u(R(t),t), \quad t > 0, \tag{12.1.8}
$$

and initial conditions

$$
\begin{aligned}
&x|_{t=0} = x_0(r), \quad y|_{t=0} = y_0(r), \quad v|_{t=0} = v_0(r) \quad \text{for } r \leq R_0 \\
&R_0 = R(0), \quad x_0 \geq 0, \quad y_0 \geq 0, \quad v_0 \geq 0, \quad x_0 + y_0 \leq \theta.
\end{aligned} \tag{12.1.9}
$$

Notice that we do not prescribe boundary conditions at $r = R(t)$ for x, y, n, because the curve $r = R(t)$ is a characteristic curve along which Eqs. (12.1.1) and (12.1.2) become ordinary differential equations.

We next introduce into the model the negative effect of the immune system on the growing cancer. Denote by $z(r, t)$ the number density of the immune cells which attack the infected tumor cells. Instead of (12.1.5) we have

$$x(r, t) + y(r, t) + n(r, t) + z(r, t) \equiv \theta,$$

and Eqs. (12.1.2) and (12.1.4) are replaced by the following equations:

$$\frac{\partial y}{\partial t} + \frac{1}{r^2}\frac{\partial}{\partial r}\left(r^2 uy\right) = \beta xv - \delta y - kzy, \qquad (12.1.10)$$

$$\frac{\partial v}{\partial t} - D\frac{1}{r^2}\frac{\partial}{\partial r}\left(r^2\frac{\partial v}{\partial r}\right) = b\delta y - \gamma v - k_0 vz$$

where

$$k \;=\; \text{immune killing rate,}$$
$$k_0 \;=\; \text{take-up rate of viruses.}$$

To model the dynamics of z we have to include a term syz where

$$s \;=\; \text{stimulation rate of infected cells,}$$

and the fact that immune cells, being cytotoxic, have a quadratic clearing rate ωz^2. Hence

$$\frac{\partial z}{\partial t} + \frac{1}{r^2}\frac{\partial}{\partial r}\left(r^2 uz\right) = syz - \omega z^2. \qquad (12.1.11)$$

Since the immune response reduces significantly the efficacy of virotherapy, biomedical researchers are experimenting with drugs that suppresses the immune system. If we denote the effect of such a drug by $P(t)z$, then we simply need to add the term $-P(t)z$ on the right-hand side of (12.1.11).

12.2. Treatment of GM-CSF

This model is based on a treatment which blocks VEGF. Macrophages are cells of the immune system which are found in abundance in the microenvironment of breast cancer and other cancers. Macrophages can be proinflammatory or anti-inflammatory, and a tumor may actually induce macrophages to become anti-inflammatory, which is helpful for tumor growth. Indeed, anti-inflammatory macrophages produce molecules that block cytotoxic T cells from killing cancer cells. In the present model we use a different fact about anti-inflammatory macrophages; namely, they produce VEGF, which stimulate angiogenesis. The drug GM-CSF makes the macrophages secrete soluble VEGF receptor (sVEGFR-1) which bonds to VEGF and neutralizes it.

Our model will include the following variables:

$$
\begin{aligned}
n &= \text{density of tumor cells,} \\
h &= \text{concentration of VEGF,} \\
e &= \text{density of endothelial cells,} \\
s &= \text{concentration of sVEGFR-1,} \\
m &= \text{density of macrophages,} \\
g(t) &= \text{concentration of GM-CSF.}
\end{aligned}
$$

We assume, for simplicity, that the density of macrophages is constant in space and time. Then, in the tumor region we have the following equations:

$$
\frac{\partial n}{\partial t} = D_n \nabla^2 n + \lambda(e)n \left(1 - \frac{n}{n_*}\right) - \mu_n n, \tag{12.2.1}
$$

$$
\frac{\partial h}{\partial t} = D_h \nabla^2 h + \lambda_1 n + \lambda_2 m - \lambda_3 hs - \mu_h h, \tag{12.2.2}
$$

$$
\frac{\partial e}{\partial t} = D_e \nabla^2 e - \operatorname{div}(\chi e \nabla h), \tag{12.2.3}
$$

$$
\frac{\partial s}{\partial t} = D_s \nabla^2 s + g(t)m - \lambda_3 hs - \mu_s s, \tag{12.2.4}
$$

where $g(t)$ represents the effect of the drug GM-CSF on the production of sVEGFR-1 by macrophages.

Notice that in (12.2.1) the growth rate $\lambda(e)$ actually stands for a nutrient-dependent parameter (we have identified endothelial cells density with nutrient concentration); thus it is natural to take

$$
\lambda(e) = \lambda_0(e - e_0), \tag{12.2.5}
$$

so that tumor cell shrinks if the endothelial (nutrient) level is below e_0.

In Eq. (12.2.2) VEGF is secreted by both cancer cells and macrophages and $\lambda_3 hs$ is the neutralization of VEGF by bonding to sVEGFR-1, and in Eq. (12.2.3) the second term on the right represents the chemotactic force of VEGF on endothelial cells.

We consider the system (12.2.1)–(12.2.5) in the radially symmetric case, in a spherical region $0 < r < L$. We take the initial conditions

$$
\begin{aligned}
n(r,0) &= (L^2 - r^2)e^{-r^2}, & h(r,0) &= 0, \\
e(r,0) &= \gamma r^2 e^{-(L-r)^2}, & s(r,0) &= 0,
\end{aligned} \tag{12.2.6}
$$

and the boundary conditions

$$
\begin{aligned}
n_r(L,t) &= 0, & h_r(L,t) &= 0, \\
e(L,t) &= \gamma L^2, & s_r(L,t) &= 0.
\end{aligned} \tag{12.2.7}
$$

As time goes on the tumor will advance toward $r = L$.

12.3. Numerical Approach for the Cancer Therapy Model

In one dimension, an easier way to solve the moving boundary problem is to map the moving domain $[0, R(t)]$ into a fixed interval $[0, 1]$ as discussed in Sect. 11.1 by introducing new coordinates:

$$\rho = \frac{r}{R(t)} \quad T = t.$$

Thus

$$\frac{\partial}{\partial t} = \frac{\partial}{\partial T} - \frac{r\dot{R}(t)}{R(t)^2}\frac{\partial}{\partial \rho} = \frac{\partial}{\partial T} - \frac{\rho\dot{R}(t)}{R(t)}\frac{\partial}{\partial \rho}, \quad \frac{\partial}{\partial r} = \frac{1}{R(t)}\frac{\partial}{\partial \rho}.$$

This leads to

$$\frac{1}{r^2}\frac{\partial}{\partial r}(r^2 u f) = u\frac{\partial f}{\partial r} + \frac{f}{r^2}\frac{\partial}{\partial r}(r^2 u) = \frac{u}{R(t)}\frac{\partial f}{\partial \rho} + \frac{\lambda x - \mu(\theta - x - y)}{\theta}f,$$

$$\frac{1}{r^2}\frac{\partial}{\partial r}\left(r^2\frac{\partial v}{\partial r}\right) = \frac{1}{\rho^2 R(t)^2}\frac{1}{R}\frac{\partial}{\partial \rho}\left(\rho^2 R(t)^2\frac{1}{R(t)}\frac{\partial v}{\partial \rho}\right) = \frac{1}{R^2}\frac{\partial^2 v}{\partial \rho^2} + \frac{2}{\rho R(t)^2}\frac{\partial v}{\partial \rho}.$$

Defining

$$F = \frac{\lambda x - \mu(\theta - x - y)}{\theta},$$

Eqs. (12.1.1)–(12.1.6) become

$$\frac{\partial x}{\partial T} + \frac{u - \rho\dot{R}(t)}{R(t)}\frac{\partial}{\partial \rho}(x) = \lambda x - \beta xv - Fx, \tag{12.3.1}$$

$$\frac{\partial y}{\partial T} + \frac{u - \rho\dot{R}(t)}{R(t)}\frac{\partial}{\partial \rho}(y) = \beta xv - \delta y - Fy, \tag{12.3.2}$$

$$n = \theta - x - y, \tag{12.3.3}$$

$$\frac{\partial v}{\partial T} - \left(\frac{\rho\dot{R}(t)}{R(t)} + \frac{2D}{\rho R(t)^2}\right)\frac{\partial v}{\partial \rho} - \frac{D}{R^2}\frac{\partial^2 v}{\partial \rho^2} = b\delta y - \gamma v, \tag{12.3.4}$$

$$\frac{1}{\rho^2 R}\frac{\partial}{\partial \rho}(\rho^2 u) = F. \tag{12.3.5}$$

The fixed and moving boundary conditions are

$$\frac{dR(t)}{dt} = u(1, t), \quad u(0, t) = 0, \quad \frac{\partial v}{\partial r}(0, t) = 0, \quad \text{and} \quad \frac{\partial v}{\partial r}(1, t) = 0.$$

PROBLEM 12.1. Compute $R(t)$ for $0 < t < T$, $T = 15$ days for the following set of parameters and initial data:

$$\lambda = 2 \cdot 10^{-2}/\text{h}, \quad \beta = \frac{7}{10} \cdot 10^{-9}\,\text{mm}^3/\text{h per virus};$$

$$\delta = \frac{1}{18}/\text{h}; \quad \mu = \frac{1}{48}/\text{h}; \quad D = 3.6 \cdot 10^{-2}\,\text{mm}^2/\text{h};$$

$$b = 50 \text{ virus/cell}; \quad \gamma = 2.5 \cdot 10^{-2}/\text{h}; \quad \theta = 10^6 \text{cells/mm}^3;$$

$$R(0) = 2\,\text{mm}; \quad x(r,0) = 0.84 \cdot 10^6 \text{cells/mm}^3; \quad y(r,0) = 0.1 \cdot 10^6 \text{cells/mm}^3;$$

$$v(r,0) = Ae^{-r^2/4}, \text{cells/mm}^3 \quad A = 5 \cdot 10^8.$$

PROBLEM 12.2. Repeat the computation of $R(t)$ with $b = 100, 200, 500$ and compare the corresponding profiles of $R(t)$.

Here we demonstrate how to solve Problem 12.1 numerically. Note that the parameters have a very wide range of values which may cause numerical problems. We thus nondimensionalized the problem via the following scaling:

$$\tilde{x} = \frac{x}{\theta}, \ \tilde{y} = \frac{y}{\theta}, \ \tilde{v} = \frac{v}{10^2\theta}, \ \tilde{T} = \frac{T}{10^2\,\text{h}}, \ \tilde{\rho} = \frac{\rho}{1\,\text{mm}}, \ \tilde{u} = 10^2 u.$$

We then obtain the following equations:

$$\frac{\partial \tilde{x}}{\partial \tilde{T}} + \frac{\tilde{u} - \rho \dot{R}(\tilde{T})}{R(\tilde{T})}\frac{\partial}{\partial \tilde{\rho}}(\tilde{x}) = \tilde{\lambda}\tilde{x} - \tilde{\beta}\tilde{x}\tilde{v} - F\tilde{x}, \qquad (12.3.6)$$

$$\frac{\partial \tilde{y}}{\partial \tilde{T}} + \frac{\tilde{u} - \rho \dot{R}(t)}{R(t)}\frac{\partial}{\partial \tilde{\rho}}(\tilde{y}) = \tilde{\beta}\tilde{x}\tilde{v} - \tilde{\delta}\tilde{y} - F\tilde{y}, \qquad (12.3.7)$$

$$\frac{\partial \tilde{v}}{\partial \tilde{T}} - \left(\frac{\rho \dot{R}(t)}{R(t)} + \frac{2\tilde{D}}{\tilde{\rho}R(t)^2}\right)\frac{\partial \tilde{v}}{\partial \tilde{\rho}} - \frac{\tilde{D}}{R^2}\frac{\partial^2 \tilde{v}}{\partial \tilde{\rho}^2} = b\tilde{\delta}\tilde{y} - \tilde{\gamma}\tilde{v}, \qquad (12.3.8)$$

$$\frac{1}{\tilde{\rho}^2 R}\frac{\partial}{\partial \tilde{\rho}}(\tilde{\rho}^2\tilde{u}) = \tilde{v}, \qquad (12.3.9)$$

where

$$F = \tilde{\lambda}\tilde{x} - \tilde{\mu}(1 - \tilde{x} - \tilde{y}).$$

The boundary conditions at $\tilde{\rho} = 0$ and $\tilde{\rho} = 1$ are

$$\frac{d\tilde{R}(\tilde{t})}{d\tilde{t}} = u(1,\tilde{t}), \quad u(0,\tilde{t}) = 0, \quad \frac{\partial v}{\partial \tilde{\rho}}(0,\tilde{t}) = 0, \quad \text{and} \quad \frac{\partial v}{\partial \tilde{\rho}}(1,\tilde{t}) = 0.$$

The parameters that we used to test our code are

$$\tilde{\lambda} = 2, \quad \tilde{\beta} = 0.07;$$

$$\tilde{\delta} = \frac{100}{18}; \quad \tilde{\mu} = \frac{100}{48}; \quad \tilde{D} = 3.6;$$

$$b = 50 \text{ virus/cell}; \quad \tilde{\gamma} = 2.5;$$

$$\tilde{R}(0) = 2; \quad x(r,0) = 0.84; \quad y(r,0) = 0.1;$$

$$v(r,0) = Ae^{-r^2/4}, \quad A = 5.$$

Let us denote the numerical solution at the nth time step by

$$(X^n, Y^n, V^n, U^n, R^n).$$

Here we use the Adams–Bashforth method to advance in time instead of the Euler method. In order to compute the solution at $(n + 1)$th step, the scheme uses the solution at both the nth and the $(n - 1)$th step. Applying the Adams–Bashforth method to

$$\frac{dR(\tilde{t})}{d\tilde{t}} = u(1, \tilde{t})$$

yields

$$R^{n+1} = R^n + \frac{\Delta t}{2} \left(3U^n - U^{n-1} \right).$$

Next the hyperbolic-type equation, e.g., (12.3.1), is solved by a leapfrog scheme

$$\frac{X_j^{n+1} - X_j^{n-1}}{2\Delta t} + A_j^n \frac{X_{j+1}^n - X_{j-1}^n}{2\Delta \rho} = \lambda X_j^n - \beta X_j^n V_j^n - F_j^n X_j^n.$$

At the two ends, the equation becomes "ODEs":

$$\frac{X_j^{n+1} - X_j^{n-1}}{2\Delta t} = \lambda X_j^n - \beta X_j^n V_j^n - F_j^n X_j^n.$$

One common problem associated with the leapfrog method is that when applied to nonlinear equations, the method often becomes unstable due to the fact that odd and even mesh points are completely decoupled. Fortunately, this mesh drifting instability can be cured by coupling the two meshes through a simple average in time:

$$X_j^n = \frac{1}{2} \left(X_j^{n+1} + X_j^{n-1} \right),$$

which still keeps the second-order accuracy of the leapfrog method. The same scheme is used to solve for Eq. (12.3.2). To compute U^{n+1}, we use the trapezoidal rule

$$\rho_{j+1}^2 U_{j+1}^{n+1} - \rho_j^2 U_j^{n+1} = \frac{R^{n+1}}{2} \Delta \rho \left[\rho_{j+1}^2 F_{j+1}^{n+1} + \rho_j^2 F_j^{n+1} \right].$$

Finally, we turn to the parabolic equation (12.3.4). We write the equation in the form

$$\frac{\partial v}{\partial T} + A_1 \frac{\partial v}{\partial \rho} + A_2 \frac{\partial^2 v}{\partial \rho^2} = RHS$$

and use the scheme

$$\frac{3V_j^{n+1} - 4V_j^n + V_j^{n-1}}{2\Delta t} + (A_1)_j^{n+1}\frac{V_{j+1}^{n+1} - V_{j-1}^{n+1}}{2\Delta\rho}$$

$$+ (A_2)_j^{n+1}\frac{V_{j+1}^{n+1} - 2V_j^{n+1} + V_{j-1}^{n+1}}{(\Delta\rho)^2} = (RHS)_j^{n+1}.$$

This discretized equation can be written in the form

$$b_j V_{j-1}^{n+1} + d_j V_j^{n+1} + a_j V_{j+1}^{n+1} = S_j, \quad j = 1, 2, \dots, J,$$

and it can be solved efficiently.

Algorithms 18–21 demonstrate the codes, and Fig. 12.1 shows the tumor radius with different burst sizes ($b = 25, 50, 100, 150, 200$) for the first 15 days. We can see that large burst size inhibits the growth of the tumor radius.

PROBLEM 12.3. Take $k = 2 \cdot 10^{-8}\,\mathrm{mm}^3/\mathrm{h}$ per immune cell, $k_0 = 10^{-8}$ mm^3/h per immune cell, $s = 5.6 \cdot 10^{-7}\,\mathrm{mm}^3/\mathrm{h}$ per infected cell, $\omega = 20 \cdot 10^{-8}\,\mathrm{mm}^3/\mathrm{h}$ per immune cell, $z(x,0) = 6 \cdot 10^4\mathrm{cells}/\mathrm{mm}^3$, and all other data as in Problem 12.1. Compute $R(t)$ for $b = 100, 200, 500$ and $t < T$, $T = 15$ days, and compare with the corresponding $R(t)$ computed in Problem 12.2. You should be able to conclude that the immune response defeats the viral therapy.

PROBLEM 12.4. Let $P(t) = 16 \cdot 10^{-2}$ for $0 < t < 72\,\mathrm{h}$ and $P(t) = 0$ for $t > 72\,\mathrm{h}$. Compute $R(t)$ under the same data as in Problem 12.3 for $b = 100, 200, 500$ and compare with the results of Problem 12.3; note the improvement achieved by the immunosuppressive drug.

PROBLEM 12.5. Repeat the computation of Problem 12.4 with different drug schedule:

$$P(t) = 8 \cdot 10^{-2} \quad \text{for } 0 < t < 72\,\mathrm{h} \quad \text{and} \quad 168 < t < 240\,\mathrm{h}$$

and $P(t) = 0$ elsewhere (so that $\int_0^{240} P(t)dt$ is the same as in Problem 12.4). Which of the two schedules yields a smaller $R(t)$?

PROBLEM 12.6. In model (12.2.1)–(12.2.7), take nondimensionalized parameters, $\lambda_0 = 5$, $n_* = 1.3$, $e_0 = 0.2$, $\mu_n = 1.2$, $D_n = D_s = 2$, $D_c = D_e = 0.01$, $\lambda_5 = 10^2$, $\lambda_2 = 3 \cdot 10^2$, $m = 1$, $\lambda_3 = 80$, $\mu_h = 30$, $\mu_s = 5$, $\chi = 10^3$, $\gamma = 1$, $L = 1$. Compute $\int_0^L r^2 c(r,t)dr$ for $0 \le t \le 10$, and draw its profile under the following two treatments:

(i) $g(t) = 1$ for $0 < t < 10$, $g(t) = 0$ elsewhere;
(ii) $g(t) = 2$ for $t_{2j} < t < t_{2j+1}$ ($j = 0, 1, 2, 3, 4$) where $t_i = i$, $g(t) = 0$ elsewhere.

Find which of the schedules yields a better result at $t = 20$; note that the total drug $\int_0^{10} g(t)dt$ is the same for both schedules.

Algorithm 18 Virotherapy algorithm-1

```
N = 48+1; % number of point in x
dx = 1/(N-1);
xx = (0:(N-1))'*dx;
dt = 0.5*(dx)^2;
Tmax = 15*24/100;
max_iter = ceil(Tmax/dt)+1;
dt = Tmax/(max_iter-1);
t = (0:(max_iter-1))*dt;
R = zeros(1,max_iter);
U = zeros(N,1);X = zeros(N,1);Y = zeros(N,1);V = zeros(N,1);
U1 = zeros(N,1);X1 = zeros(N,1);Y1 = zeros(N,1);V1 = zeros(N,1);
U2 = zeros(N,1);X2 = zeros(N,1);Y2 = zeros(N,1);V2 = zeros(N,1);
lambda = 2.0;
plot_iter = 0; col = 'bgrcmybgrcmy';
for burst = [25 50 100 150 200];
plot_iter = plot_iter+1; beta = 0.07*burst; D = 3.6; delta = 100/18;
gamma = 2.5; mu = 100/48; theta = 1;
% initialization
R(1) = 2; % mm at t=0
% first iteration
iter = 1; X(:) = 0.84; Y(:) = 0.10; a = 5; V(:) = (a*exp(-xx.^2/4));
U(1) = 0;
for i = 1:N-1
Fp = (lambda*X(i+1)-mu*(theta-X(i+1)-Y(i+1)))/theta;
Fm=(lambda*X(i)-mu*(theta-X(i)-Y(i)))/theta;
U(i+1) =
(xx(i)^2*U(i)+R(iter)/2*dx*(xx(i+1)^2*Fp+xx(i)^2*Fm))/(xx(i+1)^2);
end
R(iter+1) = R(iter)+dt*(U(N));
```

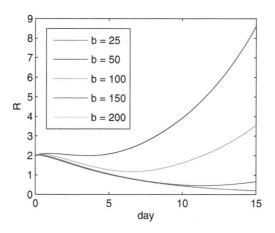

FIGURE 12.1. Tumor radius with different burst sizes for the first 15 days

---

Algorithm 19 Virotherapy algorithm-2

```
for i = 1:N
A = (U(i)-xx(i)*U(N))/R(iter+1);
F = (lambda*X(i)-mu*(theta-X(i)-Y(i)))/theta;
if i==1 | i==N
X1(i) = X(i)+ dt*(lambda*X(i)-beta*V(i)*X(i)-F*X(i));
Y1(i) = Y(i)+ dt*(beta*V(i)*X(i)-delta*Y(i)-F*Y(i));
else
X1(i) = X(i) - dt*A*(X(i)-X(i-1))/dx+...
dt*(lambda*X(i)-beta*V(i)*X(i)-F*X(i));
Y1(i) = Y(i) - dt*A*(Y(i)-Y(i-1))/dx+...
dt*(beta*V(i)*X(i)-delta*Y(i)-F*Y(i));
end
end
U1(1) = 0;
for i = 1:N-1
Fp = (lambda*X1(i+1)-mu*(theta-X1(i+1)-Y1(i+1)));
Fm = (lambda*X1(i)-mu*(theta-X1(i)-Y1(i)));
U1(i+1) =
(xx(i)^2*U1(i)+R(iter+1)/2*dx*(xx(i+1)^2*Fp+xx(i)^2*Fm))/(xx(i+1)^2);
end
A1 = -(xx.*U1(N)/R(iter+1)+2*D./(R(iter+1)^2)./xx);
A1(1) = 0;
A2 = -D./R(iter+1)^2;
```

Algorithm 19 Virotherapy algorithm-2

```
S = 2*V+2*delta*dt*Y1;
L1 = 2*dt/dx/dx*(A2)-dt/dx*A1;
D1 = (2-4*dt/dx^2*A2+2*gamma*dt)*ones(N,1);
UU1 = 2*dt/dx^2*A2+dt/dx*A1;
L1(N) = 4*dt/dx/dx*(A2)-dt/dx*A1(N);
UU1(1) = 4*dt/dx^2*A2+dt/dx*A1(1);
V1 = tridiagSolve(L1(2:N), D1(1:N), UU1(1:N-1), S);
```

Algorithm 20 Virotherapy algorithm-3

```
for iter = 2:max_iter-1
if mod(iter,20)==0
figure(1);plot(xx,X1,xx,Y1,xx,U1,xx,V1)
legend('X','Y','U','V')
title(['T = ' num2str((iter-1)*dt) ' R = ' num2str(R(iter))]); drawnow
end
R(iter+1) = R(iter)+0.5*dt*(3*U1(N)-U(N));
for i = 1:N
A = (U1(i)-xx(i)*U1(N))/R(iter+1);
F = (lambda*X1(i)-mu*(theta-X1(i)-Y1(i)));
if (i==1) | (i==N)
X2(i) = X(i)+ 2*dt*(lambda*X1(i)-beta*V1(i)*X1(i)-F*X1(i));
Y2(i) = Y(i)+ 2*dt*(beta*V1(i)*X1(i)-delta*Y1(i)-F*Y1(i));
V2(i) = (4*V1(i)-V(i)+ 2*dt*delta*Y2(i))/(3+2*gamma*dt);
else
X2(i) = X(i) - 2*dt*A*(X1(i+1)-X1(i-1))/2/dx+...
2*dt*(lambda*X1(i)-beta*V1(i)*X1(i)-F*X1(i));
Y2(i) = Y(i) - 2*dt*A*(Y1(i+1)-Y1(i-1))/2/dx+...
2*dt*(beta*V1(i)*X1(i)-delta*Y1(i)-F*Y1(i));
end
end
U2(1) = 0;
for i = 1:N-1
Fp = (lambda*X2(i+1)-mu*(theta-X2(i+1)-Y2(i+1)))/theta;
Fm = (lambda*X2(i)-mu*(theta-X2(i)-Y2(i)))/theta;
U2(i+1) =
(xx(i)^2*U2(i)+R(iter+1)/2*dx*(xx(i+1)^2*Fp+xx(i)^2*Fm))/(xx(i+1)^2);
end
A1 = -(xx.*U2(N)/R(iter+1)+2*D./(R(iter+1)^2)./xx); A1(1) = 0; A2 =
-D./R(iter+1)^2;
S = 4*V1-V+2*delta*dt*Y2; L1 = 2*dt/dx/dx*A2-dt/dx*A1;
```

Algorithm 20 Virotherapy algorithm-3

```
D1 = (3-4*dt/dx^2*A2+2*gamma*dt)*ones(N,1);
UU1 = 2*dt/dx^2*A2+dt/dx*A1; L1(N) = 4*dt/dx/dx*(A2)-dt/dx*A1(N);
UU1(1) = 4*dt/dx^2*A2+dt/dx*A1(1);
V2 = tridiagSolve(L1(2:N), D1(1:N), UU1(1:N-1), S);
X = X1;Y = Y1;U = U1;V = V1; X1 = X2;Y1 = Y2;U1 = U2;V1 = V2;
if mod(iter,50) == 0
X1 = 0.5*(X+X2); Y1 = 0.5*(Y+Y2); U1 = 0.5*(U+U2); V1 = 0.5*(V+V2);
end
end
figure(2);hold on; plot(t*100/24,R,col(plot_iter));title('R')
save(['burst_' num2str(burst) '.mat'])
end
```

Algorithm 21 Virotherapy algorithm-4

```
% tridiagSolve(l,d,u,b)
% % solves the matrix equation Ax=b via Gaussian elimination...
% where A is the tridiagonal matrix with d on the diagonal,
% l on the lower-diagonal, and u on the upper-diagonal
% d has n-elements, and l and u each have n-1
function x=tridiagSolve(l, d, u, b)
n=length(d);
% zero below the diagonal
for i=2:n
ratio=l(i-1)/d(i-1);
d(i)=d(i)-ratio*u(i-1); b
(i)=b(i)-ratio*b(i-1);
end
x=zeros(n,1);
x(n)=b(n)/d(n);
% back-substitute to solve
for i=n-1:-1:1
x(i)=(b(i)-u(i)*x(i+1))/d(i);
end
x = x;
end
```

CHAPTER 13

Granulomas

Tuberculosis remains a serious public health threat, with ten million new cases and nearly two million deaths annually. The disease is caused by *Mycobacterium tuberculosis* (Mtb). The bacterium is a respiratory pathogen usually transmitted by the cough of a person with active disease. The majority of infected humans effectively contain the disease as a latent, asymptomatic, and not contagious disease. People with active disease can be cured by treatment with several drugs over a period of at least 6 months, although there are obstacles to recovery such as access to health care, compliance, side effects, and, for some populations, efficacy of the drugs. But even recovered individuals have residual Mtb which is blocked by the same mechanism as in latent infections. It is estimated that $1/3$ of the human population are latently infected with Mtb, so an important question is what is the mechanism that blocks these latent infections from progressing to active disease.

The lungs encounter frequent challenges from inhaled particulates and microbes. Although this organ must effectively combat these invasions, it must also protect its delicate composition to ensure proper gas exchange. As a result, the lung environment is specialized to recognize and eliminate most invaders without causing excessive inflammation. However, this highly regulated inflammatory strategy can be detrimental to the host when a prompt, strong inflammatory response is needed to effectively eradicate pathogens.

Alveolar macrophages play a large role in the innate immune response of the lung. They endocytose the pathogen and digest it without causing excessive inflammation. In particular, they are able to cope with minor infection by Mtb. However, they are unable to contain a robust infection, and, in that case, the adaptive immune system is called in, which includes proinflammatory macrophages and T cells.

It is commonly known that small latent infections are concentrated in the form of granulomas. A granuloma is a collection of macrophages (and other immune cells) that surround and contain bacteria or other foreign substances. Granulomas occur not only in Mtb but also in a wide variety of other diseases including, for instance, rheumatoid arthritis, schistosomiasis, and Crohn's disease. An important biological question is then what makes granulomas stable, grow, or disappear. The answer depends of course on the specific disease.

© Springer International Publishing Switzerland 2014
A. Friedman, C.-Y. Kao, *Mathematical Modeling of Biological Processes*, Lecture Notes on Mathematical Modelling in the Life Sciences, DOI 10.1007/978-3-319-08314-8_13

In order to create a detailed, disease-specific granuloma model, one needs to consider, in addition to macrophages and bacteria, pathogen-specific cytokines, the activation state of various immune cells, and the dynamics of both extracellular and intracellular bacteria.

In this chapter, motivated by gliomas in Mtb, we develop an extremely simple model which is not disease specific and show that granulomas may shrink, grow, or be in the steady state. The model is formulated in terms of two variables:

$$
\begin{aligned}
B &= \text{bacteria external to macrophages,} \\
M &= \text{microphages uninfected by bacteria.}
\end{aligned}
$$

The equation for M is

$$
\frac{\partial M}{\partial t} + \operatorname{div}(M\mathbf{v}) = D_M \Delta M - \mu_1 M B - \alpha M, \qquad (13.0.1)
$$

where $\mu_1 B$ is the decrease rate of M caused by endocytosing bacteria and α is the death rate of M. Notice that since granulomas are dynamic, we have introduced a velocity $\mathbf{v}$ for M. We assume that the external bacteria are moving with the same velocity $\mathbf{v}$, and write the equation for B in the form

$$
\frac{\partial B}{\partial t} + \operatorname{div}(B\mathbf{v}) = D_B \Delta B - \mu_2 M B + \lambda B. \qquad (13.0.2)
$$

Here $\mu_2 M$ is the rate by which macrophages endocytose B, and λ is a growth rate of B. Actually $\lambda = \lambda(B_i, M_i)$ is a complex function of the bacteria B_i residing in infected macrophages M_i. The B_i grow in the M_i and when the M_i die naturally or by bursting under the pressure of large bacteria burdens, there emerge new external bacteria B. For simplicity we overlook this complex process and take λ to be a positive constant.

We assume that the total density of macrophages and bacteria at each point is constant and, by scaling,

$$
M + B \equiv 1. \qquad (13.0.3)
$$

Since bacteria are smaller than macrophages, their dispersion (or diffusion) coefficient D_B is large than the dispersion coefficient D_M of macrophages. We take $D_M = 1$ and then $D_B = 1 + \delta$ where $\delta > 0$. If we add Eqs. (13.0.1), (13.0.2), and use (13.0.3), we obtain

$$
\nabla \cdot \mathbf{v} = -\delta \nabla^2 M + \lambda - (\lambda + \mu + \alpha)M + \mu M^2, \qquad (13.0.4)
$$

where $\mu = \mu_1 + \mu_2$.

We shall consider only spherical granulomas with domain $\Omega(t) = \{r < R(t)\}$ which depends on time and take

$$
M = M(r, t), B = B(r, t).
$$

Writing $\mathbf{v} = v(r, t)\mathbf{e}$ where $\mathbf{e} = \mathbf{x}/r$, $|\mathbf{x}| = r$ and noting that

$$
\nabla \cdot (M\mathbf{v}) = \frac{1}{r^2} \frac{\partial}{\partial r}\left(r^2 v M\right) = \frac{1}{r^2} \frac{\partial}{\partial r}\left(r^2 v\right) M + v \frac{\partial M}{\partial r},
$$

we then obtain the following equations for M :

$$\frac{\partial M}{\partial t} + v\frac{\partial M}{\partial r} - \frac{1}{r^2}\frac{\partial}{\partial r}\left(r^2\frac{\partial M}{\partial r}\right) - M(1-M)\left[(\lambda + \mu_1 + \alpha) - \mu M\right] = 0,$$

$$0 < r < R(t), t > 0 \tag{13.0.5}$$

where

$$\frac{1}{r^2}\frac{\partial}{\partial r}\left(r^2 v\right) = -\frac{\delta}{r^2}\frac{\partial}{\partial r}\left(r^2\frac{\partial M}{\partial r}\right) + \lambda - (\lambda + \mu + \alpha)M + \mu M^2, \quad v(0,t) = 0. \tag{13.0.6}$$

We supplement the Eq. (13.0.5) with boundary condition

$$\frac{\partial M}{\partial r}(R(t),t) = \beta(1-M), \quad \beta \geq 0, \tag{13.0.7}$$

and initial conditions

$$R(0) = R_0, M(r,0) = M_0(r), 0 < M_0(r) < 1, \text{ for } 0 < r < R_0. \tag{13.0.8}$$

Finally, the free boundary $r = R(t)$ is assumed to move with the velocity v of macrophages and bacteria:

$$\frac{dR(t)}{dt} = v(R(t),t), \tag{13.0.9}$$

and using (13.0.6) we can write

$$v(r,t) = \frac{1}{r^2}\int_0^r s^2\left[-\frac{\delta}{s^2}\frac{\partial}{\partial s}\left(s^2\frac{\partial M}{\partial s}\right) + \lambda - (\lambda + \mu + \alpha)M + \mu M^2\right]ds. \tag{13.0.10}$$

A proof that there exists a unique radially symmetric solution of the granuloma model is given in [11], where some properties of the solution are also derived.

We can deduce some properties of granulomas by using the **maximum principle** for parabolic equations of the form

$$\frac{\partial u}{\partial t} - \Delta u + \sum_{i=1}^3 b_i \frac{\partial u}{\partial x_i} + cu = f$$

in a three-dimensional domain $\Omega(t)$, such as $\Omega(t) = \{x : |x| < R(t)\}$, $0 < t \leq T$. Here b_i and c are continuous functions, $f \geq 0$, $u(x,0) > 0$ in $\Omega(0)$, and

$$\alpha\frac{\partial u}{\partial \nu} + \beta u \geq 0 \text{ on the boundary of } \Omega(t) \quad \text{for} \quad 0 < t \leq T,$$

where $\alpha \geq 0$, $\beta \geq 0$, $\alpha + \beta > 0$, and ν is the outward normal. The maximum principle asserts that, under the above conditions, $u(x,t) > 0$ for all $x \in \Omega(t)$, $0 < t \leq T$.

PROBLEM 13.1. Prove that the solution of (13.0.5)–(13.0.9) satisfies the bounds:

$$0 < M(r,t) < 1,$$

for all $0 < r < R(t)$, $t > 0$. [Hint: Apply the maximum principle to the function M and to the function $u = 1 - M$.]

PROBLEM 13.2. Prove that if $\frac{\partial}{\partial r} M_0(r) > 0$ for $0 < r \leq R_0$ then $\frac{\partial}{\partial r} M(r,t) > 0$ for all $0 < r < R(t)$, $t > 0$. [Hint: Differentiate Eq. (13.0.5) in r and apply the maximum principle to the function $u = \partial M/\partial r$.]

PROBLEM 13.3. Prove that if $\lambda + \alpha < \mu_2$ and $1 - M_0(r) < \epsilon$ where $\epsilon < \frac{\mu_2 - (\lambda+\alpha)}{\mu}$, then $1 - M(r,t) < \epsilon e^{-\gamma t}$ for $0 < r < R(t)$, $t > 0$ and some $\gamma > 0$. Hence the bacteria disappear as $t \to \infty$. What is the biological implication? [Hint: The function B satisfies the equation

$$LB \equiv \frac{\partial B}{\partial t} - (1 + \delta M)\, \nabla^2 B + v\frac{\partial B}{\partial r} + (B - 1)B\,(\mu_2 - \lambda - \alpha - \mu B) = 0$$

and $w = \epsilon e^{-rt}$ satisfies $Lw > 0$ if $\gamma < c(1-\epsilon)$ where $c = \mu_2 - (\lambda+\alpha) - \mu\epsilon > 0$, and $\frac{\partial w}{\partial r} + \beta w > 0$ on $r = R(t)$. If the function $u = w - B$ satisfies $u > 0$ for $t < t_0$, $u(r,t_0) = 0$ for some $r = r_0$, $0 \leq r_0 \leq R(t_0)$, then this would contradict the maximum principle.]

PROBLEM 13.4. If $\partial M_0(r)/\partial r \geq 0$ and $\lambda > \mu_2$, $\beta = 0$, then

$$\int_0^{R(T)} r^2 B(r,T)dr \geq \int_0^{R_0} r^2 B(r,0)dr + \int_0^T \int_0^{R(T)} r^2\,(\lambda - \mu_2)\,B(r,t)drdt.$$

$$(13.0.11)$$

[Hint: Write the equation for B in the form

$$\frac{\partial B}{\partial t} - (1 + \delta)\frac{1}{r^2}\frac{\partial}{\partial r}\left(r^2\frac{\partial B}{\partial r}\right) + \frac{1}{r^2}\frac{\partial}{\partial r}\left(r^2 Bv\right) = -\mu_2 B^2 + (\lambda - \mu_2)\,B,$$

multiply it by r^2, and integrate over $0 < r < R(t)$, $0 < t < T$.]

PROBLEM 13.5. Assuming that $B(r,0) \neq 0$, derive from (13.0.11) the inequality

$$\int_0^T \int_0^{R(T)} r^2 B(r,t)drdt \geq c_0 e^{(\lambda-\mu_2)T}, T \geq 1$$

for some positive constant c_0, and, noting that $B(r,T) \leq 1$, deduce from (13.0.11) that

$$\frac{1}{3}R^3(T) \geq c_0 e^{(\lambda-\mu_2)T}.$$

From Problem 13.5 we see that if no new macrophages enter the granuloma (i.e., if $\beta = 0$) and the growth rate of external bacteria is larger than the rate of bacteria endocytosed by macrophages (i.e., $\lambda > \mu_2$), then the radius of the granuloma converges to infinity as time grows to infinity, i.e., $R(T) \to \infty$ as $T \to \infty$.

13.1. Numerical Approach to the Granulomas Model

Here we demonstrate the numerical approach [11] to solve the spherical granulomas model (13.0.5)–(13.0.10). We use the same technique in Sect. 12.3 which maps the moving domain $\Omega(t) = \{r < R(t)\}$ to a fixed unit circle $\{\rho < 1\}$ with the transformation $\rho = r/R(t)$. In the new coordinates (ρ, t), (13.0.5)–(13.0.10) becomes

$$
\frac{\partial M}{\partial t} + \left(\frac{v - \rho \dot{R}(t)}{R(t)} - \frac{2}{\rho R(t)^2} \right) \frac{\partial M}{\partial \rho} - \frac{1}{R^2} \frac{\partial^2 M}{\partial \rho^2}
$$
$$
= -\frac{2v}{\rho R} M - \frac{1}{R} \frac{\partial v}{\partial \rho} M + E(M),
\tag{13.1.1}
$$

$$
\frac{\partial}{\partial \rho} (\rho^2 v) = -\frac{\delta}{R} \frac{\partial}{\partial \rho} (\rho^2 \frac{\partial M}{\partial \rho}) + \rho^2 R F(M),
\tag{13.1.2}
$$

where $E(M) = -\mu_1 M(1 - M) - \alpha M$ and $F(M) = \lambda - (\lambda + \mu + \alpha)M + \mu M^2$ with the boundary conditions

$$
\frac{dR(t)}{dt} = v(1, t), \quad v(0, t) = 0, \quad \frac{\partial M}{\partial \rho}(0, t) = 0, \quad \text{and} \quad \frac{1}{R} \frac{\partial M}{\partial \rho}(1, t) = 1 - M.
\tag{13.1.3}
$$

Denote the numerical solution at the nth time step by

$$
(M_j^n, V_j^n, R^n)
$$

at $x_j = (j - 1)h, 1 \le j \le J$ with $(J - 1)h = 1$. We first compute V^{n+1} by the trapezoidal rule:

$$
\rho_{j+1}^2 V_{j+1}^{n+1} - \rho_j^2 V_j^{n+1} = M_j^c + \frac{R^n}{2} \Delta \rho \left[\rho_{j+1}^2 F_{j+1}^n + \rho_j^2 F_j^n \right],
$$

where

$$
M_j^c = \begin{cases} -\frac{\delta}{R} \left[\rho_{j+1}^2 \frac{M_{j+2}^n - M_j^n}{2\Delta\rho} - \rho_j^2 \frac{M_{j+1}^n - M_{j-1}^n}{2\Delta\rho} \right] & \text{for } 1 \le j \le J - 2, \\ -\frac{\delta}{R} \left[\rho_{j+1}^2 R^n \beta(1 - M_j^n) - \rho_j^2 \frac{M_{j+1}^n - M_{j-1}^n}{2\Delta\rho} \right] & \text{for } j = J - 1, \end{cases}
$$

is the central scheme approximation for the first term on the right-hand side. Euler method is used to update the radius R^n:

$$
R^{n+1} = R^n + \Delta t V^{n+1}(1, t).
$$

We write the advection–diffusion–reaction equation in the following form:

$$
\frac{\partial M}{\partial t} + A_1 \frac{\partial M}{\partial \rho} + A_2 \frac{\partial^2 M}{\partial \rho^2} = r(M, v, R)M,
$$

where

$$
A_1 = \frac{v - \rho \dot{R}(t)}{R(t)} - \frac{2}{\rho R(t)^2}, \quad A_2 = -\frac{1}{R^2},
$$

and

$$
r(M, v, R) = -(\frac{2v}{\rho R} + \frac{1}{R} \frac{\partial v}{\partial \rho}) + \mu_1(1 - M) + \alpha).
$$

We use the scheme

$$\frac{M_j^{n+1} - M_j^n}{\Delta t} + (A_1)_j^{n+1} \frac{M_{j+1}^{n+1} - M_{j-1}^{n+1}}{2\Delta\rho} + (A_2)_j^{n+1} \frac{M_{j+1}^{n+1} - 2M_j^{n+1} + M_{j-1}^{n+1}}{(\Delta\rho)^2}$$
$$= r(M_j^n, V_j^{n+1}, R^{n+1})M_j^{n+1}, \tag{13.1.4}$$

where the derivative term, $\frac{\partial v}{\partial\rho}$, at x_j in $r(M, v, R)$ is approximated by the forward difference, i.e.,

$$\frac{\partial v}{\partial\rho}\Big|_{x=x_j} \approx \frac{V_{j+1}^{n+1} - V_j^{n+1}}{\Delta\rho}.$$

This discretized equation (13.1.4) can be written in the form

$$b_j M_{j-1}^{n+1} + d_j M_j^{n+1} + a_j M_{j+1}^{n+1} = S_j, \quad j = 2, \ldots, J-1,$$

where

$$a_j = \left(\frac{(A_1)_j^{n+1}}{\Delta\rho} + 2\frac{(A_2)_j^{n+1}}{(\Delta\rho)^2}\right)\Delta t,$$

$$b_j = \left(-\frac{(A_1)_j^{n+1}}{\Delta\rho} + 2\frac{(A_2)_j^{n+1}}{(\Delta\rho)^2}\right)\Delta t,$$

$$d_j = 2 - 2\left(2\frac{(A_2)_j^{n+1}}{(\Delta\rho)^2} + r(M_j^n, V_j^{n+1}, R^{n+1})\right)\Delta t,$$

$$S_j = 2M_j^n.$$

For $j = 1$, we have the boundary condition

$$M_0^{n+1} = M_2^{n+1},$$

which implies that

$$a_1 = \left(4\frac{(A_2)_1^{n+1}}{(\Delta\rho)^2}\right)\Delta t,$$

$$d_1 = 2 - 2\left(2\frac{(A_2)_1^{n+1}}{(\Delta\rho)^2} + r(M^n)\right)\Delta t,$$

$$S_1 = 2M_1^n.$$

For $j = J$, we have the boundary condition

$$\frac{1}{R^{n+1}}\frac{M_{J+1}^{n+1} - M_{J-1}^{n+1}}{2\Delta\rho} = \beta(1 - M_j^{n+1}).$$

Thus,

$$b_J = \left(4\frac{(A_2)_j^{n+1}}{(\Delta\rho)^2}\right)\Delta t,$$

$$d_J = 2 - 2\left(2(1 + \Delta\rho\beta R^{n+1})\frac{(A_2)_j^{n+1}}{(\Delta\rho)^2} + r(M^n) + (A_1)_J^{n+1}\beta R^{n+1}\right)\Delta t,$$

$$S_J = 2M_j^n - 2\Delta t\left((A_2)_J^{n+1}\frac{2}{\Delta\rho}\beta R^{n+1} + (A_1)_J^{n+1}\beta R^{n+1}\right).$$

Note that the numerical discretization here is similar to the one mentioned in Sect. 12.3. One can simply modify Algorithms 18–21 to solve Eqs. (13.1.1)–(13.1.3).

PROBLEM 13.6. Compute $R(t)$ for $0 < t < T$, $T = 15$ days for the following set of parameters and initial data:

$$\lambda = 2 \cdot 10^{-2}/h, \quad \beta = \frac{7}{10} \cdot 10^{-9} mm^3/h \text{ per virus};$$

$$\delta = \frac{1}{18}/h; \quad \mu = \frac{1}{48}/h; \quad D = 3.6 \cdot 10^{-2} mm^2/h;$$

$$b = 50 \text{ virus/cell}; \quad \gamma = 2.5 \cdot 10^{-2}/h; \quad \theta = 10^6 \text{cells}/mm^3;$$

$$R(0) = 2mm; \quad x(r,0) = 0.84 \cdot 10^6 \text{cells}/mm^3; \quad y(r,0) = 0.1 \cdot 10^6 \text{cells}/mm^3;$$

$$v(r,0) = Ae^{-r^2/4}, \text{cells}/mm^3 \quad A = 5 \cdot 10^8.$$

PROBLEM 13.7. Repeat the computation of $R(t)$ with $b = 100, 200, 500$ and compare the corresponding profiles of $R(t)$.

Bibliography

[1] Bogacki, P., Shampine, L.F.: A 3 (2) pair of runge-kutta formulas. Appl. Math. Lett. **2**(4), 321–325 (1989)

[2] Boyce, W.E., DiPrima, R.C., Haines, C.W.: Elementary Differential Equations and Boundary Value Problems, vol. 9. Wiley, New York (1992)

[3] Brent, R.P.: Algorithms for Minimizing Without Derivatives. Prentice Hall, Englewood Cliffs (1973)

[4] Burden, R.L., Faires, J.D.: Numerical Analysis, 9th edn. Technical Report. Brooks/Cole, Pacific Grove (2011). ISBN 978-0-5387335-1-9

[5] Coddington, E.A., Levinson, N.: Theory of Ordinary Differential Equations. Tata McGraw-Hill Education, New York (1955)

[6] Craciun, G., Brown, A., Friedman, A.: A dynamical system model of neurofilament transport in axons. J. Theor. Biol. **237**(3), 316–322 (2005)

[7] Dahlquist, G., Bjorck, A.: Numerical Methods, vol. 16, pp. 37–44. Prentice-Hall, Englewood Cliffs (1974)

[8] Dormand, J.R., Prince, P.J.: A family of embedded runge-kutta formulae. J. Comput. Appl. Math. **6**(1), 19–26 (1980)

[9] Friedman, A.: Partial Differential Equations. Courier Dover Publications, New York (2011)

[10] Friedman, A., Reitich, F.: Analysis of a mathematical model for the growth of tumors. J. Math. Biol. **38**(3), 262–284 (1999)

[11] Friedman, A., Kao, C.-Y., Leander, R.: On the dynamics of radially symmetric granulomas. J. Math. Anal. Appl. **412**(2), 776–791 (2014)

[12] Gantmacher, F.R.: The Theory of Matrices, vol. 2. Taylor & Francis, London (1960)

[13] Gustafsson, B., Kreiss, H.-O., Oliger, J.: Time-Dependent Problems and Difference Methods, vol. 121. Wiley, New York (2013)

[14] Hale, J.K., Koçak, H.: Dynamics and Bifurcations. Springer, New York (1991)

[15] Iserles, A.: A First Course in the Numerical Analysis of Differential Equations, vol. 44. Cambridge University Press, Cambridge (2009)

© Springer International Publishing Switzerland 2014   147
A. Friedman, C.-Y. Kao, *Mathematical Modeling of Biological Processes*, Lecture Notes on Mathematical Modelling in the Life Sciences, DOI 10.1007/978-3-319-08314-8

[16] LeVeque, R.J., Le Veque, R.J.: Numerical Methods for Conservation Laws, vol. 132. Springer, New York (1992)

[17] MATLAB Manual: Version 2012a. Mathworks Inc. (2012)

[18] Symbolic Math Toolbox: Matlab. Mathworks Inc. (1993)

[19] Trefethen, L.N., Bau III, D.: Numerical Linear Algebra, vol. 50. SIAM, Philadelphia (1997)

[20] Van den Driessche, P., Watmough, J.: Further notes on the basic reproduction number. In: Mathematical Epidemiology, pp. 159–178. Springer, New York (2008)

[21] Wang, L., Brown, A.: Rapid intermittent movement of axonal neurofilaments observed by fluorescence photobleaching. Mol. Biol. Cell **12**(10), 3257–3267 (2001)

[22] Xu, Z., Tung, V.W.-Y.: Overexpression of neurofilament subunit m accelerates axonal transport of neurofilaments. Brain Res. **866**(1), 326–332 (2000)

Answers to Problems

2.1.   Show that

$$\frac{dy}{dt} = -ky([A] + [C]) \quad \text{and} \quad ([A] + [C])(t) = 3 \text{ for all } t > 0.$$

2.2.   Show that

$$\frac{dy}{dt} = -2k_1[A][B]^2 + 2k_2[C]$$

and

$$[A] + [C] = m, \quad 2[A] - [B] = n \text{ for all } t > 0;$$

then derive the equation

$$\frac{dy}{dt} = -k_1 y^3 - k_1 n y^2 - k_2 y + 2k_2 \left(m - \frac{n}{2}\right).$$

2.3.   Use

$$\frac{d[C_2]}{dt} = k_3[S][C_1] - (k_{-3} + k_4)[C_2] = 0$$

to deduce that the equation $d[C_1]/dt = 0$ gives

$$k_1[S][E] - (k_{-1} + k_2)[C_1] = 0$$

where $[E] = e_0 - [C_1] - [C_2]$. The first equation also yields

$$[C_2] = \frac{[C_1][S]}{K_2}, \quad K_2 = \frac{k_{-3} + k_4}{k_3},$$

and, by the second equation,

$$[C_1] = \frac{[S][E]}{K_1}, \quad K_1 = \frac{k_{-1} + k_2}{k_1}.$$

3.3.   Try two solutions:

$$(x_1, x_2) = e^{5t}(v_1, v_2), \quad (x_1, x_2) = e^{-4t}(w_1, w_2).$$

3.4.   The eigenvalues are $\lambda = 1 \pm 2i$. Try

$$x_i = \alpha_{i1} e^t \cos(2t) + \alpha_{i2} e^t \sin(2t) \quad (i = 1, 2).$$

© Springer International Publishing Switzerland 2014
A. Friedman, C.-Y. Kao, *Mathematical Modeling of Biological Processes*, Lecture Notes on Mathematical Modelling in the Life Sciences, DOI 10.1007/978-3-319-08314-8

3.5.   (a) The steady point with

$$x = \frac{1}{2}\left[(r-1) - \left((r-1)^2 + 4\right)^{1/2}\right] \quad \text{is stable}$$

and with

$$x = \frac{1}{2}\left[(r-1) + \left((r-1)^2 + 4\right)^{1/2}\right] \quad \text{is unstable.}$$

(b) The steady point is unstable.

5.3.   (i) The function $y(t) = C(t) + x(t)$ satisfies

$$y' + y = 1.$$

(ii) Show that $x'(t) < -\delta x(t)$ for some $\delta > 0$.
(iv) Show that $x' = f(x)$ where

$$f(x) = x\left(\frac{m(1-x)}{a + (1-x)} - 1\right)$$

and $f'(x) > 0$ if $x < 1 - \lambda$, $f'(x) < 0$ if $x > 1 - \lambda$.

5.4. E_1 is stable if $\lambda_1 < \lambda_2$; E_2 is unstable.
5.5. E_4 is stable if

$$k_1 < \frac{r_1}{b_1}, \ k_2 < \frac{r_2}{b_2}.$$

5.6.   (i) $(0, \bar{y}, 0)$ is unstable.
(ii) The steady point is stable.

5.7.   (ii)

$$\bar{u} = \frac{\delta}{1 - \delta\beta}, \quad \bar{v} = (1 + \beta\bar{u})\, r\, (\bar{u} - \alpha)\, (1 - \bar{u}).$$

If $\beta \sim 0$ then the Jacobian is approximately

$$\begin{pmatrix} r\,(1 - 2\bar{u} + \alpha) & -\bar{u} \\ 1 & 0 \end{pmatrix}.$$

5.9.   Use the Routh-Hurwitz theorem and the approximations $x_0 \sim B$, $z_0 \sim B$ where $B = \gamma A/(\gamma + \delta)$ is large.

6.1. $p = 0$ is a transcritical bifurcation; $p = -\frac{1}{4}$ is a saddle point bifurcation.

6.4.   The characteristic polynomial is

$$\lambda^2 + a\lambda + b = 0, \quad a = -p\,(2p - 1), \quad b = 4p^2\,(1 - p).$$

6.5.   The characteristic polynomial is

$$\lambda^2 + a\lambda + b = 0$$

where

$$a = \frac{1 - 4p}{4 - p}, \quad b = \frac{6}{4 - p}\left(2 - \frac{6}{4 - p}\right).$$

7.1.   The function $z = x + y$ satisfies $\frac{dz}{dt} = z + \epsilon z^2$. Deduce that $z(t) = (1 + \epsilon)\, e^{-t} - \epsilon$, and $T_\epsilon = \ln\left((1 + \epsilon)/\epsilon\right)$.

7.2.   The two nullclines intersect at a point $(\bar{x}, \bar{y})$ where

$$\lambda\bar{x} - 100 = \bar{x}\,(\bar{x} - 1)\,(2 - \bar{x})\,,$$

and the cubic $y = x\,(x - 1)\,(2 - x)$ is an increasing function if $a < x < b$, where a, b can be computed explicitly. The slope λ must be such the intersection point $(\bar{x}, \bar{y})$ satisfies: $a < \bar{x} < b$. Hence $B < \lambda < A$ where

$$A = \frac{1}{a}\,[100 - a\,(a - 1)\,(2 - a)]\,, \quad B = \frac{1}{b}\,[100 - b\,(b - 1)\,(2 - b)]\,.$$

8.1. $u(x,t) = \cos^2(x - 2t)$ if $x > 2t$, $u = 1$ if $x < 2t$.

8.2. $u(x,t) = \sin(xe^{-t}) + t^3/3 + t$.

10.1.   The right-hand side of (10.0.1) is equal to

$$\alpha\left(x - \frac{1}{2}\Delta x\right) u\,(x - \Delta x) + \alpha\left(x + \frac{1}{2}\Delta x\right) u\left(x + \frac{1}{2}\Delta x\right)$$

$$-\left[\alpha\left(x - \frac{1}{2}\Delta x\right) + \alpha\left(x + \frac{1}{2}\Delta x\right)\right] u(x)$$

where, for brevity, we write $u(x)$ instead of $u(x,t)$. By rearrangement, the last expression is equal to

$$\Delta x\left[\alpha\left(x - \tfrac{1}{2}\Delta x\right)\frac{u(x-\Delta x)-u(x)}{\Delta x} + \alpha\left(x + \tfrac{1}{2}\Delta x\right)\frac{u(x+\Delta x)-u(x)}{\Delta x}\right]$$

$$= (\Delta x)^2\left\{\frac{-\alpha\left(x-\frac{1}{2}\Delta x\right)\frac{\partial u}{\partial x}\left(x-\frac{1}{2}\Delta x\right)+\alpha\left(x+\frac{1}{2}\Delta x\right)\frac{\partial u}{\partial x}\left(x+\frac{1}{2}\Delta x\right)}{\Delta x} + O(|\Delta x|)\right\}$$

$$= (\Delta x)^2\left\{\frac{\partial}{\partial x}\left(\alpha\,(x)\,\frac{\partial u(x)}{\partial x}\right) + O\,(|\Delta x|)\right\}.$$

10.2.   (ii)

$$u = \begin{cases} \text{const} \cdot \exp\left[\frac{w^{\alpha+1}}{D(\alpha+1)}\right] & \text{if} \quad \alpha \neq -1, \\ \text{const} \cdot w^{1/D} & \text{if} \quad \alpha = -1. \end{cases}$$

13.2.   The function $w = \frac{\partial M}{\partial r}$ satisfies

$$\frac{\partial w}{\partial t} + (1 + \delta M)\,\Delta w + a\frac{\partial w}{\partial r} + bw = 0$$

for some coefficients a, b, and $w \geq 0$ on the boundary.

13.5.   The function

$$w(T) = \int_0^T \int_0^{R(T)} r^2 B(r,t)dr dt$$

satisfies

$$w'(T) \geq (\lambda - \mu_2)\,w(T) \quad \text{for all } T \geq 0$$

and $w'(0) > 0$, so that $w(t_0) > 0$ for some small t_0. Deduce that

$$w(T) \geq \text{const} \cdot e^{(\lambda-\mu_2)(T-t_0)} \quad \text{if} \quad T > t_0.$$

Index

© Springer International Publishing Switzerland 2014
A. Friedman, C.-Y. Kao, *Mathematical Modeling of Biological
Processes*, Lecture Notes on Mathematical Modelling in the Life
Sciences, DOI 10.1007/978-3-319-08314-8